KB040038

SAT II
Chemistry

Final Summary & Chapter Tests
5 Final Full-length Tests

머리말

SAT II Chemistry는 국내외 국제학교 및 외국어고등학교 등 유학을 준비하는 학생들이 가장 많이 준비하는 Subject Test 시험 중 하나이다. 특히 의·치의·약학 계열을 포함한 이공계 진학을 목표로 하는 학생들에게는 필수적인 요소일 만큼 입시 및 입시 후의 과정과 연계되어 매우 중요한 과목이라고 할 수 있다.

SAT II Chemistry는 AP Chemistry 및 그 이상의 심화과정을 위해 반드시 정리가 필요한 필수 화학 개념을 다루는 과목이다. 자연과학 과목은 단순히 개념을 들었다고 하여 문제가 바로 풀리지 않는다. 이는 문제를 해결하기 위해서는 보다 정교한 문제풀이의 논리가 필요하기 때문이다. 10여년 이상 많은 학생들과 수업하면서 "개념은 이해가 되지만 문제 해결이 잘되지 않는다."는 상담을 많이 받았다. 시험을 준비하는 과정에서 문제풀이는 개념을 정확하게 정리하고 체계화시킴으로써 문제 해결의 전체적인 방법론을 익히는 것을 목표로 한다. 이 과정이 잘 이루어지기 위해서는 양질의 문제를 많이 해결하면서 부족한 부분을 보완하여야 한다. 하지만 안타깝게도 유학을 준비하는 학생들이 참고할만한 다양한 문제를 제공하는 교재가 많지 않다.

이 교재는 현장 강의를 수강하는 학생들뿐만 아니라 시험을 준비하는 많은 학생들에게 도움을 주고자 출간하게 되었다. 현장에서 학생들과 함께 해결하였던 문제들 중 도움이 될 수 있는 양질의 문제들을 엄선하여 구성하였다. 아무쪼록 이 교재가 학생들이 체계적으로 시험을 준비할 수 있는 좋은 가이드가 되기를 진심으로 바란다.

끝으로 이 교재가 출판될 수 있도록 많은 도움을 주신 마스터프렙의 권주근 대표님께 깊은 감사의 말을 전한다.

2020년 12월

저자 소개

임성택(Simon Lim)

(현) 마스터프렙 화학 대표강사
(전) 한영외국어고등학교 AP Chemistry 출강
(전) 대원외고, 민사고 및 미국 명문 보딩스쿨 학생 지도

10여년 이상 현장 및 마스터프렙의 인터넷 동영상 강의를 통해 수많은 학생들의 대학 진학에 힘써왔다. 함께 했던 국내외 국제학교 및 외국어고등학교 학생들은 현재 화학, 생명 등 다양한 관련분야에서 학업을 진행 중에 있거나 사회에 진출하여 자신의 역량을 펼치고 있다.

전공 박사 출신으로 교과부터 화학올림피아드의 내용까지 통합적이고 연계성이 있는 강의 커리큘럼으로 학생들이 화학을 보다 체계적으로 학습할 수 있도록 양질의 컨텐츠를 제공하기 위해 꾸준히 노력해왔다.

교재의 특징

이 교재는 SAT II Chemistry 시험을 준비하는 학생들이 단원별 내용을 정확하게 빠르게 리뷰하고 관련문제를 해결함으로써 개념의 문제화를 통해 실전감각을 익히는 chapter test 파트와 5 세트의 final full test 파트로 구성되어 있다.

각 파트의 구성은 학생들이 문제를 통해 출제 가능한 내용에 대한 정확한 이해와 문제풀이 방법론을 익힐 수 있는 양질의 문제를 포함하고 있다. 특히 보기를 비롯한 문제의 텍스트도 학습효과가 나타날 수 있도록 하여 한 문제를 통해서도 많은 개념과 내용을 정리할 수 있도록 구성하였다.

이러한 교재의 구성을 통해 학생들은 제한된 시간에 연계된 내용을 체계적으로 정리할 수 있고 문제를 통해 개념을 다시 확인하고 문제풀이에 활용된 개념을 더욱 공고히 이해하고 활용하는 데 최적화된 결과를 얻을 수 있을 것이다.

이 교재는 공개된 기출문제가 다루었던 개념과 함께 현장에서 학생들이 가장 효과적으로 시험을 준비할 수 있었던 결과가 반영되어 이 교재를 활용하여 시험을 준비하는 많은 학생들에게 많은 도움이 될 수 있을 것이라 기대해 볼 수 있다.

CONTENTS

[Chapter Review and Test]

01 Matter and Measurement ·· 8

02 Atom, Molecule and Ion ·· 17

03 Electron Configuration ·· 25

04 Periodic Table and Periodicity ·································· 32

05 Chemical Bond ·· 41

06 Gas ·· 49

07 Solid, Liquid, and Phase Change ·································· 55

08 Solution ·· 63

09 Thermochemistry ·· 73

10 Chemical Kinetics ·· 81

11 Chemical Equilibrium ·· 90

12 Acid and Base ·· 100

13 Electrochemistry ·· 109

14 Organic Chemistry ·· 119

15 Nuclear Chemistry ·· 123

[Full Test]

CHEMISTRY TEST No.1 ·· 123

CHEMISTRY TEST No.2 ·· 136

CHEMISTRY TEST No.3 ·· 150

CHEMISTRY TEST No.4 ·· 163

CHEMISTRY TEST No.5 ·· 176

[Answer Key]

Chapter Review and Test ·· 198

Full Test ·· 203

Periodic Table of the Elements

1	2	3	4	5	6	7	8	9	10	11	12	13	14	15	16	17	18
1 H 1.008																	2 He 4.00
3 Li 6.94	4 Be 9.01											5 B 10.81	6 C 12.01	7 N 14.01	8 O 16.00	9 F 19.00	10 Ne 20.18
11 Na 22.99	12 Mg 24.30											13 Al 26.98	14 Si 28.09	15 P 30.97	16 S 32.06	17 Cl 35.45	18 Ar 39.95
19 K 39.10	20 Ca 40.08	21 Sc 44.98	22 Ti 47.87	23 V 50.94	24 Cr 52.00	25 Mn 54.94	26 Fe 55.85	27 Co 58.93	28 Ni 58.69	29 Cu 63.55	30 Zn 65.38	31 Ga 69.72	32 Ge 72.63	33 As 74.92	34 Se 78.97	35 Br 79.90	36 Kr 83.80
37 Rb 85.47	38 Sr 87.62	39 Y 88.91	40 Zr 91.22	41 Nb 92.91	42 Mo 95.95	43 Tc	44 Ru 101.07	45 Rh 102.91	46 Pd 106.42	47 Ag 107.87	48 Cd 112.41	49 In 114.82	50 Sn 118.71	51 Sb 121.76	52 Te 127.60	53 I 126.90	54 Xe 131.29
55 Cs 132.91	56 Ba 137.33	57–71 *	72 Hf 178.49	73 Ta 180.95	74 W 183.84	75 Re 186.21	76 Os 190.23	77 Ir 192.22	78 Pt 195.08	79 Au 196.97	80 Hg 200.59	81 Tl 204.38	82 Pb 207.2	83 Bi 208.98	84 Po	85 At	86 Rn
87 Fr	88 Ra	89–103	104 Rf	105 Db	106 Sg	107 Bh	108 Hs	109 Mt	110 Ds	111 Rg	112 Cn	113 Uut	114 Uuq	115 Uup	116 Uuh	117 Uus	118 Uuo

Lanthanides

57 La 138.9	58 Ce 140.12	59 Pr 140.91	60 Nd 144.24	61 Pm	62 Sm 150.36	63 Eu 151.97	64 Gd 157.25	65 Tb 158.93	66 Dy 162.50	67 Ho 164.93	68 Er 167.26	69 Tm 168.93	70 Yb 173.05	71 Lu 174.97

Actinides

89 Ac	90 Th 232.04	91 Pa 231.04	92 U 238.03	93 Np	94 Pu	95 Am	96 Cm	97 Bk	98 Cf	99 Es	100 Fm	101 Md	102 No	103 Lr

Chapter Review Test

01 Matter and Measurement

1.1 Classification of Matter

1.1.1 Pure substance

- Element

  Element can not be separated into simpler substances by chemical means. **ex** He, O_2, Fe,..

- Compound

  Compound is composed of two or more elements chemically united in fixed proportions.

  ex H_2O, NaCl, Fe_2O_3,...

1.1.2 Mixture

- Homogeneous mixture

  The composition of homogeneous mixture is the same throughout the mixture

  ex brass(Zn/cu), KNO_3(aq), Air,...

- Heterogeneous mixture

  The composition of heterogeneous mixture is not the same throughout the mixture(can be separated into two or more homogeneous mixture)

1.2 Properties of Matter

1.2.1 Physical and chemical properties

- Physical properties

  Physical properties can be measured without changing the chemical composition

  ex phase change, dissolving,...

• Chemical properties

Chemical properties are observed and measured in chemical reaction (ex) rusting of metal. neutralization,...

1.2.2 Quantitative and qualitative properties

• Quantitative properties

Quantitative properties can be measured with numbers (ex) melting point, density,...

• Qualitative properties

Qualitative properties can be described but not measured (ex) color, ductility,...

1.3 Separation of Mixture

1.3.1 Using the difference of boiling point

• Simple distillation

The process that depends on difference in the volatility (ex) $H_2O(l)$ from $NaCl(aq)$,...

• Fractional distillation

The process that depends on difference of boiling point

(ex) $C_2H_5OH(l)$ from $C_2H_5OH(aq)$, gasoline from crude oil,...

1.3.2 Using the difference of density

• Separatory funnel

Separatory funnel is the equipment used in liquid-liquid extraction to separate in two immiscible solvent phases densities

• Floating and settling

The process is used to separate the mixture by whether the materials is floating or settling (ex) gold form mineral

1.3.3 Using the solubility

• Filtration

The process can be used when a mixture is composed of solid and liquid (ex) $NaCl(aq)$ mixed with sand

• Extraction

Solubility difference in two different immiscible liquids (ex) extraction from $I_2(aq)$ to hexane

• Fractional crystallization

The process is a method of refining substances based on differences in their solubility

ex mixture of KNO_3 and KCl

1.3.3   Using the affinity to mobile and stationary phases

• Chromatography

The process depends on different affinities for the mobile phase to stationary phase of component and different rate of flow (rate of flow = distance the component flow/distance to the finish line)

1.4   Measurement and Significant Figure

1.4.1   Measurement

• Base unit

Physical quantity	Name of unit	Abbreviation(Symbol)
Mass	kilogram	kg
Length	meter	m
Temperature	kelvin	K
Time	second	s
Amount of substance	mole	mol
Electric current	ampere	A
Luminous intensity	candela	cd

• Derived unit

Physical quantity	Name of unit	Abbreviation(Symbol)
Volume	cubic meter	m^3
Density	kilogram per cubic meter	kg/m^3

• Temperature : Celsius(℃), Kelvin(K), and Fahrenheit(°F)
• Volume : buret, pipet, graduated cylinder, volumetric flask

1.4.2   Accuracy and precision

• Accuracy

  How close a measurement comes to the actual dimension or true value of whatever is measured

• Precision

  Reproducibility of the measurement

1.4.3   Significant figure

Significant figure is all the digits that can be known precisely plus the last digit that must be estimated in measurement and calculation

 ## Rules for counting significant figure

1. Nonzero
   Nonzero integers always count as significant figures
2. Zero
   ① Zeros that precede all the nonzero do not count as significant figure
   ② Zeros between nonzero digits(captive zero) always count as significant figure
   ③ Trailing zeros(zeros at the right end of the number) significant only if the number contains a decimal point
3. Exact numbers
   Numbers that are not obtained using measuring devices but are determined by counting

• Scientific notation

Scientific notation means the expression of value using power of 10 using the significant figure range from least 1 to less 10

• Significant figure in calculation

Addition and subtraction	The number of significant figures in the result is the same as the number of decimal places as the least precise measurement used in calculation.
Multiplication and division	The number of significant figures in the result is the same as the number in the fewest significant figures used in the calculation.
Rules of rounding	If the digit to be removed A. is less than 5, the preceding digit stays same B. is equal to or greater than 5, the preceding digit is increased by 1 $\begin{array}{r} 1.234 \\ + 1325.1 \\ \hline 1326.334 \end{array}$ 1326.3 $\begin{array}{r} 3.26 \\ \times 4.2 \\ \hline 13.692 \end{array}$ 14

• Matric Prefixes

	Factor	Abbreviation		Factor	Abbreviation
tera	10^{12}	T	centi	10^{-2}	c
giga	10^{9}	G	milli	10^{-3}	m
mega	10^{6}	M	micro	10^{-6}	μ
kilo	10^{3}	k	nano	10^{-9}	n
heco	10^{2}	h	pico	10^{-12}	p
deca	10^{1}	da	femto	10^{-15}	f
deci	10^{-1}	d	atto	10^{-18}	a

Chapter Review Questions **1. Matter and Measurement**

[01-04] **Refer to the following experimental methods**

(A) fractional distillation   (B) fractional crystallization

(C) filtration   (D) chromatography

---

01 Technique for separation of precipitate from a liquid mixture using the funnel

02 Technique for separation of the liquid mixture using the difference of boiling point

03 Technique for separation of the mixture using the solubility of solid solute

04 Technique for separation of the mixture using the difference of affinity to mobile phase

05 The burning of CH_4(methane) is chemical change

> Because

There are rearrangement of chemical bond and chemical properties between reactants and products are different

06 The density of 100g of water is doubled compared with that of 50g of water at constant temperature

> Because

The density means the mass of unit volume

07 If there is no systemic error, measurement using pipette is more accurate than graduated cylinder

> Because

Accuracy means how close a measurement comes to the actual dimension or true value of whatever is measured

08 Two samples are massed using different balances. When the mass of sample 1 is 6.2890g and the mass of sample 2 is 2.55g. which of the following is correct?

I. The balance used for sample 1 is precision

II. The number of significant figure of sample 1 is 4

III. The number of significant figure of sum of mass 1 and 2 is 3

(A) I only   (B) II only   (C) I, II only   (D) I, III only   (E) II, III only

09 Which of the following is not correct?

(A) Element can not be decomposed into several substance by chemical methods

(B) Aqueous solution is homogeneous mixture

(C) The phase of solution is always liquid

(D) Compound can be decomposed into simper substances by chemical means

(E) Alloy is homogeneous mixture

10 Which of the following is not chemical change?

(A) dissolving NaCl in water

(B) burning a piece of wood

(C) ozone absorbing ultraviolet light

(D) dissolving Na metal in water

(E) rusting of metal

11 Which of the following physical property of H_2O is correct?

(A) $H_2O(l)$ can be decomposed into $H_2(g)$ and $O_2(g)$

(B) $H_2O(l)$ can react with alkali metal to form a basic solution

(C) The density of $H_2O(l)$ is always constant

(D) $H_2O(s)$ is less dense than $H_2O(l)$

(E) The heat of vaporization of $H_2O(l)$ is relatively smaller than that of other molecules with similar molar mass

12 Which of the following explain is not for the chemical reaction?

(A) Total number of atoms in both reactants and products is the same

(B) There is rearrangement of atoms and new substance is formed

(C) Sum of mole number of reactants is the same as that of products

(D) The chemical and physical properties between reactants and products are different

(E) Compound can be formed from combining of elements in definite proportion

13 Sucrose is the chemical name for the sugar we consume. Its solubility at $20\,°C$ is 204g/100g water, and at $100\,°C$ is 487g/100g water. A solution is prepared by mixing 139g of sugar in 33.0g of water at $100\,°C$ Which of the following is not correct?

(A) More Sugar can be dissolved in higher temperature than lower temperature

(B) 139g of sugar can be dissolve in 33.0g of water completely at $100\,°C$

(C) The sucrose crystalline can be formed when (B) solution is cooled to $20\,°C$

(D) Solution is saturated solution

(E) The percent concentration of saturated solution can not be 100%

14 Which of the following is not correct for the chemical change?

(A) The combustion of hydrocarbon
(B) $NH_3(g)$ is formed from $N_2(g)$ and $H_2(g)$
(C) The formation of $CO_2(g)$ from dry ice
(D) The formation of $CO_2(g)$ from heating of $CaCO_3(s)$
(E) The formation of $NH_4Cl(s)$ from $NH_3(g)$ and $HCl(g)$

15 Which of the following is not correct for the chemical change?

(A) Water($H_2O(l)$) can be decomposed into hydrogen($H_2(g)$) and oxygen($O_2(g)$)
(B) The distillation of salt water($NaCl(aq)$) produce water vapor($H_2O(g)$)
(C) The heating of baking powder($NaHCO_3(s)$) can produce carbon dioxide($CO_2(g)$) and sodium carbonate($Na_2CO_3(s)$)
(D) The combustion of ethanol(C_2H_5OH) can produce water($H_2O(l)$) and carbon dioxide($CO_2(g)$)
(E) When calcium carbonate $CaCO_3(s)$ is added to $H_2CO_3(aq)$, $Ca(HCO_3)_2(aq)$ is formed

16 Which of the following is correct for the temperature conversion set?

	Celsius(℃)	Fahrenheit(°F)	Kelvin(K)
(A)	265	536	536
(B)	265	536	536
(C)	305	581	578
(D)	305	578	581
(E)	335	603	670

17 What is the density of 25.00g of tablet with 12mL?

(A) 2g/mL   (B) 2.1g/mL   (C) 2.08g/mL
(D) 2.083g/mL   (E) 2.0833g/mL

18 An object having a mass of 18.25 grams is placed into a graduated cylinder containing water. The level of the water rose from 20.4mL to 26.7mL. Which of the following is correct for the density of the object with significant figures?

(A) 3g/mL   (B) 2.9g/mL   (C) 2.90g/mL
(D) 2.897g/mL   (E) 2.8968g/mL

19 Which of the following is not correct?

	Calculation	Number of significant figures
(A)	$50.1 - 3.25$	3
(B)	$42.0 + 2.217$	3
(C)	2.5×2.5	2
(D)	2.11×4.2	3
(E)	$5.67 \div 2.3$	2

20 Which of the following is not correct for the prefix?

(A) 10^{-3} mili
(B) 10^{-2} centi
(C) 10^{3} kilo
(D) 10^{6} mega
(E) 10^{9} nano

02 Atom, Molecule and Ion

2.1   **Classical Atomic Model**

2.1.1   Dalton's atomic model

• Element is made up of tiny particles called atom
• The atoms of a given element are identical
• Compounds are formed when atoms of different elements combine with each other
• Chemical reactions involve reorganization of the atoms not changed in a chemical reaction

2.1.2   Thomson's cathode ray experiment

• Cathode ray is bent by both electric and magnetic field
• Rays are composed of negatively charged particles called electrons
• Flume pudding model of atom

2.1.3   Rutherford's alpha particle scattering experiment

• α-particle, positively charged particle, deflected by the gold foil
• A small, dense core with positive charge nucleus and electrons are outside and most of the atom is empty space.
• Nuclear model of atom

2.2   Particles of Atom

2.2.1   Proton, neutron and electron

Particles	Mass	Charge	
		Coulomb	Relative Charge
Electron	9.109×10^{-31} kg	-1.6022×10^{-19} C	-1
Proton	1.673×10^{-27} kg	$+1.6022 \times 10^{-19}$ C	$+1$
Neutron	1.675×10^{-27} kg	None	None

2.2.2   Atomic number, isotope and average atomic mass

• Atomic number(z) is the number of proton

• Mass number is the sum of protons plus neutrons

• Isotopes are atoms that have the same number of protons, but a different number of neutrons. The chemical properties of isotopes are the same but different how stable their nucleus is

2.3   Fundamental chemical laws in chemical reaction

2.3.1   Chemical law for atom

• Law of conservation of mass
  Mass is neither created nor destroyed in chemical reaction

• Law of definite proportion
  A given compound always contains exactly the same proportion of element by mass

• Law of multiple proportion
  When two elements form series of compounds, the ratio of the masses of the second element that combine with 1g of the first element can always be reduced to small whole number

2.3.2   Chemical law for molecule

• Gay-Lussac's law(Law of combining volumes)
  At the same temperature and pressure, the volume ratio between reacting and producing gas can be expressed in simple whole number

- Avogadro's law

  Equal volumes of gases at same temperature and pressure contain the same number of molecules, regardless of chemical and physical properties of molecules, particularly, 22.4L of gas contains $6.02 \times 10^{23}(N_A)$ molecules at $0°C$, 1atm,

- Avogadro's Number(N_A) and mole concept

  1mol is the same as the number of 12.00g of ^{12}C atom and molar mass is the mass in grams of 1mole of the substance(atom, molecule, ion, ionic compound,...)

2.4 Chemical Formula

2.4.1 Chemical formula

- Molecular formula represents the exact numbers of all elements in the smallest unit of substance
- Empirical formula represents the simplest whole number ration of each element in compound
- Structural formula represents the relative positions and bond in molecule
- Condensed structural formula represent the functional group of the molecule which belongs to

2.4.2 Finding the empirical formula and molecular formula

- % composition

  ex H_2O $H \% \dfrac{2.00gH}{18.00g} \times 100(\%) = 11.19\% \, H$ $O \% \dfrac{16.00gH}{18.00g} \times 100(\%) = 88.81\% \, O$

- Empirical formula from % composition

  ex 40.0% C, 6.6% H, 54.4% O   the mole ratio C : H : O = 1 : 2 : 1   empirical formula CH_2O

- Experimental determination of empirical formula(Liebig analysis)

  ex 22mg of CO_2, 18mg of H_2O the mole ration C : H = 1 : 4   empirical formula CH_4

- The relationship between empirical formula and molecular formula

  (empirical formula)n = molecular formula   n = (molecular formula mass)/(empirical formula mass)

  ex molecular formula mass of molecule with empirical formula, CH_2O is 60g/mol   molecular formula $C_2H_4O_2$

2.4.3 Ion

- Cation is formed from losing the electron(s)
- Anion is formed from gaining the electrons(s)

2.4.4 Ionic compound

• Type I ionic compound is ionic compound composed of representative elements or polyatomic ion

  ex NaCl(sodium chloride), $Ca(NO_3)_2$(calcium nitrate), K_3PO_4(potassium phosphate), NH_4NO_3(ammonium nitrate),...

• Type II ionic compound is ionic compound which contains transition element

  ex Fe_2O_3(iron(III) oxide), Cu_2O(copper(I) oxide), $Co(NO_3)_2$(cobalt(II) nitrate), Cr_2O_3,...(chromium(III) oxide)

• Type III molecule is composed of representative element

  ex CO(carbon monoxide), N_2O(dinitrogen monoxide), NO_2(nitrogen dioxide), HF(hydrogen fluoride),...

Element		Cation		Element		Anion	
Na	sodium	Na^+	sodium ion	O	oxygen	O^{2-}	oxide ion
K	potassium	K^+	potassium ion	S	sulfur	S^{2-}	sulfide ion
Mg	magnesium	Mg^{2+}	magnesium ion	F	fluorine	F^-	fluoride ion
Al	aluminium	Al^{3+}	aluminium ion	Cl	chlorine	Cl^-	chloride ion
Cu	copper	Cu^+	copper(I) ion	Br	bromine	Br^-	bromide ion
Cu	copper	Cu^{2+}	copper(II) ion	I	iodine	I^-	iodide ion
Fe	iron	Fe^{2+}	iron(II) ion	N	nitrogen	N^{3-}	nitride ion
Fe	iron	Fe^{3+}	iron(III) ion	P	phosphorus	P^{3-}	phosphide ion

Polyatomic ion	Name	Polyatomic ion	Name
NH_4^+	ammonium ion	CN^-	cyanide ion
OH^-	hydroxide ion	PO_4^{3-}	phosphate ion
NO_3^-	nitrate ion	ClO_4^-	perchlorate ion
SO_4^{2-}	sulfate ion	ClO_3^-	chlorate ion
HSO_4^-	hydrogen sulfate ion	ClO_2^-	chlorite ion
CO_3^{2-}	carbonate ion	ClO^-	hypochlorite ion
HCO_3^-	hydrogen carbonate ion	MnO_4^-	permanganate ion

Chapter Review Questions   **2. Atom, Molecule and Ion**

[21-22] **Refer to the following**

(A) F_2 and Cl_2

(B) ^{12}C and ^{14}C

(C) C_2H_5OH(ethanol) and CH_3OCH_3(dimethyl ether)

(D) He and Ar

(E) C(graphite) and C(diamond)

21 are relationship of isomer

22 are relationship allotrope

23 Atoms of different elements can have the same mass number

           Because

    The atoms of each element have a characteristic number of protons in the nucleus

24 At STP($0°C$, 1atm), 22.4L of He(g) will have the same volume as one mole of H_2(g)

           Because

    One mole or 22.4L of any gas at STP will have the same mass.

25 Which of the following is not for the Dalton's Atomic Theory?

(A) Matter is composed of tiny particle called atom

(B) Compounds are made up of combination of atoms in definite proportion

(C) All atoms given element are the same

(D) A chemical reaction involves the rearrangement of atoms

(E) Protons, neutrons, and electrons are sub-particles of atom

26 Which of the following is not for the Thomson's Cathode ray experiment?

(A) Thomson discovered the sub-atomic particle called electrons

(B) Electrons have negative charged particle

(C) Electrons can be deflected in magnetic field but can not be deflected in electric field

(D) Thomson proposed the plume pudding model

(E) Thomson found the charge to mass ratio of electron but could not calculate the each value of mass and charge of electron

27 Which of the following is not for the Rutherford's Gold Foil Experiment?

(A) Rutherford discovered the existence of proton

(B) Alpha particles have positive charge

(C) The mass of atom is concentrated in the center of atom

(D) Most of the alpha particles pass through the space of atom

(E) Some of the alpha particles reflected in center of atom, which explain the existence of dense and positive charged particles

28 Average atomic mass of Cl(chlorine) is 35.5. There are two types of Cl(chlorine) : ^{35}Cl(Atomic Mass : 34.97) and ^{37}Cl(Atomic Mass : 36.97) Which of the following is not correct?

(A) ^{35}Cl and ^{37}Cl are isotope

(B) ^{35}Cl is more abundant than ^{37}Cl

(C) Two types of Cl have the same chemical properties

(D) Two types of Cl have the same number of protons

(E) Two type of Cl have the same number of neutrons

29 Which of the following is the correct for the Na−23 ion at ground state?

	proton	electron neutron	neutron
(A)	11	11	11
(B)	11	10	12
(C)	11	11	12
(D)	12	12	12
(E)	12	13	12

30 Which of the following is correct for the 0.5mol of CO_2 gas at STP?

	Number of CO_2 molecules	Number of atoms	Volume
(A)	6.02×10^{23}	6.02×10^{23}	11.2L
(B)	6.02×10^{23}	9.03×10^{23}	22.4L
(C)	3.01×10^{23}	6.02×10^{23}	11.2L
(D)	3.01×10^{23}	9.03×10^{23}	11.2L
(E)	3.01×10^{23}	6.02×10^{23}	22.4L

$N_2(g) + 3H_2(g) \rightarrow 2NH_3(g)$

31 At STP, he volume of Hydrogen gas required to produce 22.4L of ammonia gas is

(A) 11.2L   (B) 22.4L   (C) 33.6L   (D) 44.8L   (E) 56.0L

32 $$...CaCO_3(s) + ...HCl(aq) \rightarrow ...CaCl_2(aq) + ...H_2O(l) + ...CO_2(g)$$

Which of the following is not correct for the above reaction?(The molar mass of $CaCO_3$: 100g/mol, C : 12g/mol, H : 1g/mol, O : 16g/mol)

(A) The sum of each coefficient as whole number is 6
(B) When 1mol of $CaCO_3(s)$ react with excess $HCl(aq)$, 44.8L of $CO_2(g)$ is produced at STP
(C) When 1mol of $CO_2(g)$ is produced 1mol of $CaCl_2(aq)$ is formed
(D) The mol ratio of $CaCO_3$ to HCl is 1 : 2
(E) When 100g of $CaCO_3$ is reacted with excess $HCl(aq)$ 44g of $CO_2(g)$ is produced

33 $$C_3H_8(g) + 5O_2(g) \rightarrow 3CO_2(g) + 4H_2O(l)$$

The combustion of propane, $C_3H_8(g)$, proceeds according to the equation above. How many grams of water will be formed in the complete combustion of 44.0g of propane?(The molar mass of C : 12g/mol, H : 1g/mol, O : 16g/mol)

(A) 4.50g   (B) 18.0g   (C) 44.0g   (D) 72.0g   (E) 176g

34 $$...CH_4(g) + ...O_2(g) \rightarrow ...CO_2(g) + ...H_2O(l)$$

Which of the following is not correct for the above reaction?

(A) The sum of coefficient as whole number is 6
(B) At STP, 22.4L of $CO_2(g)$ is produced when 1mol of $CH_4(g)$ is completely reacted with excess $O_2(g)$
(C) The mol number of reacted $O_2(g)$ is the same as the produce number of $CO_2(g)$
(D) At STP, 1mol of $CH_4(g)$ reacted with 44.8L of $O_2(g)$
(E) The sum of mol number of reactants is the same as that of products

35 A compound is composed of 52.2% C, 13.0% H, 34.8% O. What is the empirical formula?(The molar mass of C : 12g/mol, H : 1g/mol, O : 16g/mol)

(A) CH_2O (B) C_2H_4O (C) C_2H_6O (D) $C_5H_{13}O_3$ (E) $C_5H_{12}O_2$

36 Which of the following is correct for the % composition of CH_3NH_2? The molar mass of C : 12g/mol, H : 1g/mol, N : 14g/mol)

	C(%)	H(%)	N(%)
(A)	16.1	5.0	78.9
(B)	45.2	5.0	78.9
(C)	38.7	10.0	73.9
(D)	38.7	11.0	50.3
(E)	38.7	16.1	45.2

37 Which of the following is not correct for the 4g sample of $H_2(g)$ and 4g sample of $O_2(g)$ and the below reaction?(The molar mass of H : 1g/mol, O : 16g/mol)

$$2H_2(g) + O_2(g) \rightarrow 2H_2O(g)$$

(A) The sample of $H_2(g)$ occupy 44.8L at STP
(B) The mol number of $H_2(g)$ is twice than that of $O_2(g)$ before the reaction
(C) The limiting reactant is $O_2(g)$
(D) The mol number of produced gas is 0.25mol
(E) The total mol number of produced and remained gas is 2mol

38 Which of the following is correct for the volume of CO_2 gas produced when 11.2L of CO gas reacts with 11.2L of O_2 gas and which gas is remained at constant temperature and pressure?

$$2CO(g) + O_2(g) \rightarrow 2CO_2(g)$$

	$CO_2(g)$ produced	Remaining gas
(A)	5.6L	5.6L of $O_2(g)$
(B)	11.2L	5.6L of $O_2(g)$
(C)	22.4L	5.6L of $CO_2(g)$
(D)	33.6L	11.2L of $O_2(g)$
(E)	44.8L	11.2L of $CO_2(g)$

39 $$...Fe_2O_3(s) + ...CO(g) \rightarrow ...CO_2(g) + ...Fe(s)$$

Which of the following is not for the above reaction?

(A) The sum of coefficients as whole number is 9
(B) When 1mol of Fe_2O_3 is reacted with 1mol of CO, limiting reactant is CO(g)
(C) 67.2L of CO is needed to react with 1mol of Fe_2O_3 at STP
(D) When 1mol of Fe_2O_3 reacted with CO completely, the mol number of produced gas is 3mol
(E) When 1mol of Fe_2O_3 reacted with excess CO and 0.5mol of Fe is produced, the % yield is 50%

40 Which of the following is correct for the name of compounds?

(A) $CaCO_3$ calcium carbonate Na_2O sodium monoxide N_2O dinitrogen oxide
(B) $FeCl_3$ iron chloride KCl potassium chloride NO_2 nitrogen dioxide
(C) Cu_2O copper (II) oxide CaF_2 calcium fluoride CO carbon monoxide
(D) Fe_2O_3 iron(III) oxide NaCl sodium chloride N_2O_5 nitrogen pentoxide
(E) FeO iron(II) oxide NaBr sodium bromide CO_2 carbon dioxide

03 Electron Configuration

3.1 Bohr's Atomic Model

3.1.1 Electromagnetic radiation

• Electromagnetic radiation is the energy moving way through space

• Electromagnetic radiation has the electric filed and magnetic field and two components have the same wavelength and frequency

• Light is the electromagnetic radiation and has the continuous spectrum

3.1.2 Bohr's atomic model

• The energy level of atom is not continuous, quantized and discrete energy level of electron

• Absorption and emission spectra : the light with specific wavelength can be absorbed or released when electron transits form certain energy level to another level

 Series of line spectrum of hydrogen

Lyman series $n=1 \leftrightarrow n=2,3,4,...$ UV(ultraviolet)
Balmer series $n=2 \leftrightarrow n=3,4,5,...$ Vis(visible)
Paschen series $n=3 \leftrightarrow n=4,5,6,...$ IR(infrared)
Brakett series $n=4 \leftrightarrow n=5,6,7,...$ IR(infrared)
Pfund series $n=5 \leftrightarrow n=6,7,8,...$ IR(infrared)

3.2   Modern Atomic Model

3.2.1   Heisenberg's uncertainty principle

It is impossible to know simultaneously both the momentum (mass times velocity) and the position of a particle with certainty

3.2.2   Schrödinger and Quantum number

• Principal quantum number(n) : the energy level of orbital. $n = 1, 2, 3, ...$

• Angular momentum quantum number(ℓ) : the shape of orbital. $\ell = 0, 1, 2, ... n-1$

• Magnetic quantum number(m_ℓ) : the orientation of orbital in space. $m_\ell = -\ell, ... 0, ... +\ell$

• Spin quantum number(m_s) : the spin of electron. $+\dfrac{1}{2}$ or $-\dfrac{1}{2}$

n	ℓ	orbital	m_ℓ	m_s
1	0	1s	0	
2	0	2s	0	
	1	2p	$-1, 0, +1$	
3	0	3s	0	
	1	3p	$-1, 0, +1$	$\pm\dfrac{1}{2}$
	2	3d	$-2, -1, 0, +1, +2$	
4	0	4s	0	
	1	4p	$-1, 0, +1$	
	2	4d	$-2, -1, 0, +1, +2$	
	3	4f	$-3, -2, -1, 0, +1, +2, +3$	

3.2.3   Rules for electron configuration

• Aufbau principle

  The electrons are filled one after the other in low energy orbital.

• Hund's rule

  The most stable electron configuration is the one with the greatest number of parallel spin in same energy level

• Pauli's exclusion principle

  If two electron have the same n, ℓ, m_ℓ, the spin of each electron which occupy the same orbital is opposite

3.2.4 Electron configuration of representative elements and transition elements

• Electron configuration of representative elements

Element	Electron configuration	Element	Electron configuration
H	$1s^1$	Na	$1s^2 2s^2 2p^6 3s^1$
He	$1s^2$	Mg	$1s^2 2s^2 2p^6 3s^2$
Li	$1s^2 2s^1$	Al	$1s^2 2s^2 2p^6 3s^2 3p^1$
Be	$1s^2 2s^2$	Si	$1s^2 2s^2 2p^6 3s^2 3p^2$
B	$1s^2 2s^2 2p^1$	P	$1s^2 2s^2 2p^6 3s^2 3p^3$
C	$1s^2 2s^2 2p^2$	S	$1s^2 2s^2 2p^6 3s^2 3p^4$
N	$1s^2 2s^2 2p^3$	Cl	$1s^2 2s^2 2p^6 3s^2 3p^5$
O	$1s^2 2s^2 2p^4$	Ar	$1s^2 2s^2 2p^6 3s^2 3p^6$
F	$1s^2 2s^2 2p^5$	K	$1s^2 2s^2 2p^6 3s^2 3p^6 4s^1$
Ne	$1s^2 2s^2 2p^6$	Ca	$1s^2 2s^2 2p^6 3s^2 3p^6 4s^2$

• Electron configuration of 3d transition elements

Element	Long form	Short form	Element	Long form	Short form
Sc (scandium)	$1s^2 2s^2 2p^6 3s^2 3p^6 4s^2 3d^1$	$[Ar]4s^2 3d^1$	Fe (Iron)	$1s^2 2s^2 2p^6 3s^2 3p^6 4s^2 3d^6$	$[Ar]4s^2 3d^6$
Ti (Titanium)	$1s^2 2s^2 2p^6 3s^2 3p^6 4s^2 3d^2$	$[Ar]4s^2 3d^2$	Co (Cobalt)	$1s^2 2s^2 2p^6 3s^2 3p^6 4s^2 3d^7$	$[Ar]4s^2 3d^7$
V (Vanadium)	$1s^2 2s^2 2p^6 3s^2 3p^6 4s^2 3d^3$	$[Ar]4s^2 3d^3$	Ni (Nickel)	$1s^2 2s^2 2p^6 3s^2 3p^6 4s^2 3d^8$	$[Ar]4s^2 3d^8$
Cr (Chromium)	$1s^2 2s^2 2p^6 3s^2 3p^6 4s^1 3d^5$	$[Ar]4s^1 3d^5$	Cu (Cooper)	$1s^2 2s^2 2p^6 3s^2 3p^6 4s^1 3d^{10}$	$[Ar]4s^1 3d^{10}$
Mn (Manganese)	$1s^2 2s^2 2p^6 3s^2 3p^6 4s^2 3d^5$	$[Ar]4s^2 3d^5$	Zn (Zinc)	$1s^2 2s^2 2p^6 3s^2 3p^6 4s^2 3d^{10}$	$[Ar]4s^2 3d^{10}$

• Paramagnetic and diamagnetic : substance with electron configuration contains unpaired electron attracted in magnetic filed and substance with electron configuration do not contain unpaired electron is repelled in magnetic field

Chapter Review Questions ▶ **3. Electron Configuration**

[41-45] **Refer to the following**

(A) Aufbau principle

(B) Hund's rule

(C) Heisenberg uncertainty principle

(D) Pauli's exclusion principle

(E) Bohr's model of hydrogen

---

41 Energy level of atom is quantized (incontinuous)

42 No electrons with same 4 quantum number can occupy the same orbital

43 The momentum and position of electron can not be known simultaneously

44 The most stable arrangement of electrons in sub-shells is the one with the greatest number of parallel spins

45 Electron will enter the orbital in order of increasing energy

---

[46-48] **Refer to the following elements**

(A) O   (B) Ar   (C) S

(D) Ti   (E) U

---

46 has the ground state of electron configuration $1s^2 2s^2 2p^6 3s^2 3p^4$

47 has the partially filled f orbital in ground electron configuration

48 has the most stable electron configuration at ground state

49 When electron absorb the light energy, it can be excited state and occupies the higher energy level

Because

We can not know momentum and position of electron simultaneously

50 The ground state of electron configuration of phosphorus is $1s^22s^22p^63s^23d^3$

Because

Phosphorus has d orbital

51 Which of the following is not correct?

(A) Energy level of electron is quantized
(B) A specific light energy is absorbed when electron transits from ground state to excite state
(C) Hydrogen spectra is line spectrum
(D) Line spectrum is continuous spectra
(E) Energy of line spectrum is inversely proportional to wavelength of light absorbed or emitted

52 Which of the following is not correct for the Bohr's atomic model?

(A) When electron is excited form $n=1$ to $n=2$, UV light is absorbed.
(B) The energy of Balmer series is smaller than that of Lyman series.
(C) Energy level of hydrogen atom is quantized.
(D) Energy difference between $n=2$ and $n=3$ is smaller than that of between $n=3$ and $n=4$
(E) Energy needed to transit the electron from $n=1$ to $n=2$ is larger than that of $n=2$ to $n=6$

53 Which of the following is not correct for the electron configuration?

(A) Electron configuration of Cu is $[Ar]4s^13d^{10}$
(B) Electron configuration of Fe^{3+} is $[Ar]3d^5$
(C) $1s^22s^22p^53s^23p^5$ is the electron configuration of Cl
(D) If the electron configuration of neutral atom is $1s^22s^22p^63s^23p^63d^2$, this is excited state of Ca
(E) There are 3 unpaired electrons(same spins) in d orbital of ground state of Cobalt atom

54 Which of the following is not for the ground state electron configuration of $1s^22s^22p^63s^23p^6$

(A) Electron configuration can be ground state electron configuration of argon
(B) Electron configuration can be ground state electron configuration of potassium ion
(C) If the electron configuration is $+1$ ion, the neutral atom is sodium
(D) If the electron configuration is -2 ion, this ion have same number of electrons of Ca^{2+}
(E) If the electron configuration is negative charged ion, this ion can form ionic bond with sodium ion

55 Which of the followings has partially filled electron configuration of d orbital

(A) Ca   (B) F   (C) Ne   (D) Co   (E) S

56 Which of the following is not correct for the magnetism of elements in magnetic field?

(A) Substances that contain net unpaired spins are attracted by a magnet and show paramagnetic properties.

(B) Substances that do not contain net unpaired spins are not attracted by a magnet and show diamagnetic properties.

(C) Elements of 2A and 8A show paramagnetic properties

(D) Oxygen shows paramagnetic property

(E) Zn shows diamagnetic property

57 Which of the following is not for the set of quantum number and orbital?

	n	ℓ	m_ℓ	m_s	orbital
(A)	1	0	0	+ 1/2	1s
(B)	2	0	0	− 1/2	2s
(C)	2	1	− 1	− 1/2	2s
(D)	3	2	0	− 1/2	3d
(E)	3	1	+ 1	+ 1/2	3p

58 When 6 electrons are filled in the 3d orbital in the ground state electron configuration, how many electrons are paired in the 3d orbital?

(A) 1   (B) 2   (C) 3   (D) 4   (E) 5

59 Which of the following is correct for the combination of quantum number of electron with highest energy of Zn?

	n	ℓ	m_ℓ	m_s
(A)	2	1	0	+ 1/2
(B)	3	2	+ 2	+ 1/2
(C)	3	1	− 3	+ 1/2
(D)	4	3	− 2	− 1/2
(E)	4	2	+ 1	+ 1/2

60 Which neutral element has the following electron configuration?

$1s^22s^22p^63s^23p^64s^13d^5$

(A) Sc   (B) V   (C) Cr   (D) Co   (E) Cu

04   Periodic Table and Periodicity

4.1   Periodic Table

4.1.1   Period and group

- Period is the horizontal row in periodic table. Elements with same period have the same valence shell at ground state.
- Group is the vertical columns in periodic table. Element with same group shows similar chemical properties because each element has the same valence electron

Main group elements (Representative elements)	Elements of group 1(1A),2(2A), 13(3A), 14(4A), 15(5A), 16(6A), 17(7A), 18(8A). electrons of s orbital or p orbital are valence electron at ground state
Transition elements (d-block transition elements)	Transition elements have partially d or f subshell. Elements of group from group 3 to group 12 are d-block transition elements(period 4~6) partially or fully filled d orbital at ground state. particularly, transition metals are colored.
Lanthanides and Actinides (f-block transition elements)	Lanthanides(period 6) and Actinides(period 7) are f-block transition elements. They have partially or fully filled f orbital. Lanthanides are the 14 elements after lanthanum. Actinides are the 14 elements after actinium

4.1.2   Metal, nonmetal, and metalloid

- Metal is a good conductor of heat and electricity. Metal is malleable and ductile. Metallic properties are comprised from metallic bond particularly, free electrons (ex) Na, Ca, Al, Cr, Fe,...
- Nonmetal is usually a poor conductor of heat and electricity. Nonmetal is brittle (ex) C, N, O, F, He,...
- Metalloid has properties that are intermediate between those of metal and nonmetal (ex) B, Si, Ge, As, Sb,...

4.2 Periodicity

- **Effective nuclear force(charge) : Z_{eff}**

Effective nuclear force is the nuclear force experienced by an electron concerned with attractive force from nucleus and repulsive force from other electrons. As atomic number increases, Z_{eff} increases in same period. As atomic number increases, Z_{eff} increases in same group.

> − Atomic number increases in same period → Effective nuclear force(Z_{eff}) increases
> − Atomic number increases in same group → Effective nuclear force(Z_{eff}) increases

- **Atomic radius**

As atomic number increases, atomic radius decreases at same period because Z_{eff} is increasing as increasing of atomic number. As atomic number increase, atomic radius increases at same group because valence electron shell is increasing as increasing of atomic number

> − Atomic number increases in same period → Atomic radius decreases(Z_{eff} increases)
> − Atomic number increases in same group → Atomic radius increases(Valence shell number increases)

- **Ionic radius**

Cations(positively charged ions) are smaller than those of neutral atoms because Z_{eff} for outermost electron increase as valence electron(s) shell is removed. Anions(negatively charge ions) larger than those of neutral atoms because Z_{eff} for outermost electron decrease as added electron(s) increase the shielding effect

> − Cation is smaller than that of neutral atom(Z_{eff} increases from the decreases of outermost electron shell number)
> − Anion is larger than that of neutral atom(Z_{eff} decrease from the increases of electron repulsion, shielding effect)
> − The isoelectronic are atoms or ions which have the same number of electron. At ground state, as the proton number increases, the radius of isoelectronic decreases

- **(The first) Ionization energy : $A(g) \rightarrow A^+(g) + e^-$ (Always endothermic)**

Ionization energy is the minimum energy needed to remove an electron from a gaseous atom. As the atomic number increases, the first ionization energy increases at same period, because Z_{eff} incases form left to right across a period. As the atomic number increase the first ionization energy decreases at same group because valence electron shell is increasing as increasing of atomic number

> − Atomic number increases in same period → The first ionization energy increases(Z_{eff} increases)
> − Atomic number increases in same group → The first ionization energy decreases(Valence shell number increases)

Exception of first ionization energy

> • The first ionization energy of element 2A is larger than that of 3A at the same period because the energy of np is higher than that of ns (n is principal quantum number)
> • The first ionization energy of element 5A is larger than that of 6A at the same period because electron pair repulsion of np orbital in 6A

• **Successive ionization energy**

Successive ionization energy is just the progressive removal of electrons for an atom. As the electron removed form atom the more energy is required to remove electron successively. Core electrons are bounded to the nucleus than valence electrons, so there is largely increase in ionization energy when the shell of electron removed decreases.

ex Mg : 1st(737.7kJ/mol), 2nd(1,450.7kJ/mol), 3rd(7,732.7kJ/mol), 4th(10,542.5)

> – From the successive ionization energy, the number of valence electrons can be known : If the number of valence electrons is (n), there is largely jump in (n+1)th ionization energy compare with (n)th ionization energy

• **Electron affinity**

Electron affinity is the energy released when a electron added to neutral gaseous atom. The higher the electron affinity, the more energy is released. As the atomic number increases, the electron affinity increases at same period, because Z_{eff} incases form left to right across. As the atomic number increase the first ionization energy decreases at same group because valence electron shell is increasing as increasing of atomic number.

> – Atomic number increases in same period → Electron affinity increases(Z_{eff} increases)
> – Atomic number increases in same group → Electron affinity decreases(Valence shell number increases)

• **Electronegativity**

Electronegativity is the ability to attract the bonding pair of electrons in a chemical bond. Electronegativities of Linus Pauling have no unit

4.3 Properties of Groups

4.3.1 Alkali metal

Order of chemical reactivity(First ionization energy)					
Element	Li	Na	K	Rb	Cs
Ionization energy	520	496	419	403	376

React with water(H_2O)

$2Na(s) + 2H_2O(l) \rightarrow 2NaOH(aq) + H_2(g)$ $2M(s) + 2H_2O(l) \rightarrow 2MOH(aq) + H_2(g)$ (M : Alkali metal)
The color of phenolphthalein changes from colorless to pink

React with oxygen(O_2)

$4Na(s) + O_2(g) \rightarrow 2Na_2O(s)$ $4M(s) + O_2(g) \rightarrow 2M_2O(s)$ (M : Alkali metal)
Alkali metal is good reducing agent in chemical reaction

Flame reaction

Element	Li	Na	K	Rb	Cs
Flame color	Red	Yellow	Purple	Red	Blue

React with halogen and form a ionic compound

$2Na(s) + Cl_2(g) \rightarrow 2NaCl(s)$
• Na(s) is oxidized and reducing agent and $Cl_2(g)$ is reduced and oxidizing agent
• Na^+ and Cl^- are bonded through electrostatic force(coulomb force)

Order of physical property

Element	Li	Na	K	Rb	Cs
Melting Point(K)	453.5	370.8	336.2	312.0	301.5
Boiling Point(K)	1620	1154.4	1038.5	961.0	978.0
Density(g/cm^3)	0.53	0.97	0.86	1.53	1.90

4.3.1 Alkaline earth metal

Alkaline earth metals are similar with the alkali metals in chemical reaction

React with water(H_2O)

$Ca(s) + 2H_2O(l) \rightarrow Ca(OH)_2(aq) + H_2(g)$ $M(s) + 2H_2O(l) \rightarrow 2M(OH)_2(aq) + H_2(g)$ (M : Alkaline earth metal)
The color of phenolphthalein changes from colorless to pink

React with oxygen(O_2)

$2Ca(s) + O_2(g) \rightarrow 2CaO(s)$ $2M(s) + O_2(g) \rightarrow 2MO(s)$ (M : Alkaline earth metal)
Alkali metal is good reducing agent in chemical reaction

React with halogen and form a ionic compound
$Ca(s) + 2Cl_2(g) \rightarrow CaCl_2(s)$
• $Ca(s)$ is oxidized and reducing agent and $Cl_2(g)$ is reduced and oxidizing agent
• Ca^+ and Cl^- are bonded through electrostatic force(coulomb force)

Flame reaction of alkaline earth metals

Element	Mg	Ca	Sr	Ba	Ra
Flame color	Briliant-white	Brick-red	Crimson	Apple green	Crimson-red

4.3.3 Halogen

Element	M.P(°C)	B.P(°C)	First IE(kJ/mol)	EA(kJ/mol)	Density(g/cm)	Phase(room temp.)
F_2	-218.6	-188.1	1680.6	322.6	1.5	Gas
Cl_2	-101.0	-34	1255.7	348.5	1.7	Gas
Br_2	-7.3	59.5	1142.7	324.7	3.2	Liquid
I_2	113.6	185.2	1008.7	295.5	4.0	Solid

4.3.4 Oxide

• Oxide of metals is base anhydride and reacts with acid

$$Na_2O(s) + 2HNO_3(aq) \rightarrow 2NaNO_3(aq) + H_2O(l)$$

• Oxide of nonmetal is acid anhydride and reacts with base

$$CO_2(g) + Ca(OH)_2(aq) \rightarrow CaCO_3(s) + H_2O(l)$$

• Amphoteric oxides can react with acid or base ex Al_2O_3, ZnO,...

Chapter Review Questions | **4. Periodic Table and Periodicity**

[61-62] Refer to the following electron configuration

(A) $1s^2 2s^2 2p^6 3s^2 3p^2$
(B) $1s^2 2s^2 2p^6 3s^2 3p^4$
(C) $1s^2 2s^2 2p^6 3s^2 3p^5$
(D) $1s^2 2s^2 2p^6 3s^2 3p^6$
(E) $1s^2 2s^2 2p^6 3s^2 3p^6 4s^1$

---

61 has the largest first ionization energy

62 its oxide is basic

.

[63-64] Refer to the following properties

(A) (First) Ionization energy
(B) Successive ionization energy
(C) Electron Affinity
(D) Electronegativity
(E) Ionic radius

---

63 The energy required of positive ion is larger than that of its neutral atom and from this we can predict the valence electrons of neutral atom

64 The character that pull the electron pair in covalent bond

65 The first ionization energy of Li(lithium) is smaller than that of Na(sodium)

Because

Effective nuclear force of Li(lithium) is smaller than that of Na(sodium)

66 Oxygen has a smaller first ionization energy than fluorine.

Because

Oxygen has a higher Z_{eff} value than fluorine.

67 The third ionization energy of B is higher than that of Be.

$\boxed{\text{Because}}$

The third electron to be removed from B and Be comes from the same principal energy level.

68 Which of these give(s) a correct trend in ionization energy?

I. $Al < Si < P < Cl$
II. $Be < Mg < Ca < Sr$
III. $I < Br < Cl < F$
IV. $Na^+ < Mg^{2+} < Al^{3+} < Si^{4+}$

(A) I only   (B) III only   (C) I, and II only   (D) I, and IV only   (E) I, III, and IV only

69 Which of the following properties increase with increasing of atomic number in 1A group?

I. (The first) Ionization energy
II. Atomic radius
III. Density
IV. Electron affinity

(A) I only   (B) III only   (C) I, and II only   (D) II, and III only (E) I, III, and IV only

70 Which of the following is correct for the transition metals?

I. Transition metal can exist at various oxidation states
II. Electrons are partially filled in 4d orbital in transition metal in period 4 at ground state
III. Transition element can form a complex ion in its aqueous solution and shows specific color

(A) I only   (B) II only   (C) III only   (D) I, II only   (E) I, III only

71 Which of the following is correct for atomic number and scientist of modern periodic table?

	Atomic number based	Scientist
(A)	Its atomic radius	Rutherford
(B)	Its ionic radius	Thomson
(C)	Its atomic mass	Mendeleev
(D)	Its number of protons	Moseley
(E)	Its number of neutrons	Millikan

72 Which of the following is correct for the elements with limited chemical reactivity?

(A) Alkali metal   (B) Alkaline earth metal   (C) Transition metal
(D) Halogen   (E) Noble gas

73 Which of the following is correct for the halogens?

(A) Atomic radius : F > Cl > Br > I
(B) They exist as solid in nature
(C) They can act as oxidizing agent in chemical reaction with alkali metal
(D) Electronegativity : F < Cl < Br < I
(E) The first ionization energy F < Cl < Br < I

74 Which of the following is not correct for the periodic table?

(A) The horizontal rows of elements in the periodic table are called periods and elements with same periods have same electron shell numbers.
(B) Elements in the same vertical columns in the periodic table are called group and elements of same group have similar chemical properties.
(C) The group 8A exists as monoatomic molecule at STP(0℃, 1atm) and they are very reactive with metal ion
(D) The group 7A are good oxidizing agent than group 8A
(E) Generally nonmetallic properties increase from left to light in the periodic table

75 Which of the following elements has the greatest electronegativity?

(A) O   (B) F   (C) N   (D) Ne   (E) Mg

76 Which of the following is correct for the (first) ionization energy?

(A) Energy change associated with the addition an electron to a gaseous neutral atoms
(B) Energy required to remove a electron from a gaseous neutral atoms in its ground state
(C) Generally ionization energy of metals is larger than nonmetal
(D) Ionization energy of noble gas is the smallest in same period because of its stable electron configuration
(E) Nonmetals have negative ionization energy because nonmetal ions is more stable than neutral atoms

77 The below is the table of successive ionization energy

(kJ/mol)

1st	2nd	3rd	4th	5th	6th
1087	2353	4621	6223	37831	47277

This element is

(A) B   (B) C   (C) N   (D) O   (E) F

78 Which of the following is not correct for the periodic table?

(A) Generally metallic property of main group elements decrease from left to light in the periodic table
(B) Generally metallic property of main group elements increase from top to bottom in the periodic table
(C) Group 1 elements are alkali metal because they can react with water and form alkaline solution
(D) Alkali metal ion can form ionic compound with ions of halogen(halide)
(E) Alkali metal can oxidize nonmetal

79 Which of the following is not correct for the periodicity?

(A) Radius $K^+ > Ca^{2+}$
(B) First ionization energy $Na < Mg$
(C) Chemical reactivity with water $Li > Na$
(D) Chemical reactivity as oxidizing agent $F_2 > Cl_2$
(E) Successive ionization energy of Na $IE_1 \ll IE_2 < IE_3 < \cdots$

80 Which of the following is not correct?

(A) Metal is good conductor of electricity
(B) Nonmetal is easy to be processed because of its malleability and ductility
(C) Metalloid has both metallic and nonmetallic properties
(D) Effective nuclear force of K(Potassium) is larger than Na(sodium) because proton number of K is larger than Na
(E) The melting point of Na is higher than K

05   Chemical Bond

5.1   Ionic Bond

5.1.1   The formation of ionic bond

• Ionic bond is formed through electrostatic force between cation and anion

• Lattice energy is the energy change when gaseous cation and anion are bonded to form a ionic crystalline and vice versa. It can be calculated by Coulomb's law

> As the charge of each ion increases, the lattice energy increases. When the charge of each ion is the same, as the radius of ion decreases, the lattice energy decreases

• Generally, as the lattice energy increases, the melting point of ionic compound increases

5.1.2   The Characteristics of Ionic Compound

• Ionic compounds may be soluble or insoluble in water (solubility)

• Generally, melting and boiling point are higher than molecules

• Aqueous solution and molten conduct electricity

5.2   Covalent bond

5.2.1   The formation of covalent bond

• The covalent bond is formed between two atom through sharing of electrons

• The bond length is the distance between two atoms in a molecule

• Generally, as the bond length decreases, bond energy increases

Bond	Bond length, pm	Bond energy(enthalpy), kJ/mol
C − C	154	347
C = C	134	614
C≡C	120	839
C − O	143	358
C = O	123	745
C≡O	113	1072
C − N	143	305
C = N	138	615
C≡N	116	891

5.2.2 Sigma(σ) bond and pi(π) bond

• The electron density is high on the axis of bond in sigma bond

• The electron density is high above and below of the bond in pi bond

5.2.3 Lewis structure, molecular geometry(VSEPR : Valence Shell Electron Pair Repulsion) and hybrid orbital

• The way to write Lewis structure

> − Sum of valence electrons from all the atoms
> − Use a pair of electrons to form a bond between each pair of bound atoms
> − Arrange the remaining electrons to satisfy the duet rule for hydrogen and the octet rule for the second-row elements

• Coordinate covalent bond(dative bond) is the covalent bond where one of the atom donate two electrons as bonding electrons

ex NH_3 donates the lone pair electrons as bonding electrons when it is bonded with BF_3

• Steric number(SN) of the atom is the sum of sigma bond plus lone pair

• Molecular geometry is affected by the electron pair repulsion(minimize the electron pair repulsion)

> lone pair − lone pair(LP − LP) > lone pair − Bonding pair(LP − BP) > Bonding pair − Bonding pair(BP − BP)

• Steric number is the same as the number of hybrid orbitals

Electron pair			Molecular geometry			Example		Hybrid orbital
Total	Bonding	Lone						
2	2	0	Linear	Linear AX$_2$		BeF_2	F—Be—F	sp
3	3	0	Trigonal planar	Trigonal planner AX$_3$		BF_3		sp^2
	2	1		Bent(or angular) AX$_2$	—Lone pair	SO_2		
4	4	0	Tetrahedral	Tetrahedral AX$_4$		CH_4		sp3
	3	1		Trigonal pyramidal AX$_3$		NH_3		
	2	2		Bent(or angular) AX$_2$		H_2O		

5.2.4 Resonance structure

Resonance structure is two or more Lewis structure which describes the delocalization of electrons in a molecule. The real structure is hybrid of several resonance structures. **ex** benzene(C_6H_6), ozone(O_3),...

5.3   Polarity of Molecule

5.3.1 Polarity of bond

• Nonpolar covalent bond is the bond between two identical atoms
• Polar covalent bond is the bond between different atoms, so bonding electron pair is pulled to more electronegative atom

5.3.2 Polarity of molecule

The polarity of a molecule is related to its geometric symmetry
• Polar molecule has permanent dipole moment.

• Nonpolar molecule don't has permanent dipole moment. This is the case of a non-polar covalent bond or a polar bond, but the dipole of the bond is canceled.

Types of Covalent bond	Symmetry	Polarity of Molecule	Example
Polar Covalent Bond	Asymmetrical	Polar molecule	NH_3, H_2O, $CHCl_3$,...
	Symmetrical	Nonpolar Molecule	CO_2, BF_3, CH_4,...
Nonpolar Covalent Bond	Symmetrical		N_2, O_2, F_2
	Asymmetrical	Polar molecule	O_3

Chapter Review Questions   5. Chemical Bond

[81-83] **Refer to the following molecules**

(A) H_2
(B) N_2
(C) NH_3
(D) CO
(E) CO_2

---

81 Nonpolar molecule with polar covalent bond

82 has only one unshared pair of electron with sp^3 hybrid orbital

83 has one triple bond with permanent dipole

84 Atoms of the same element combine covalently rather than by ionic attraction

Because

There is no electronegativity difference between same atoms.

85 Benzene molecule is stable

Because

Electrons in pi bond are delocalized in molecule in resonance structure

86 Molecules that contain a polar bond are not necessarily polar compounds.

Because

If polar bonds in a molecule are symmetrically arranged, then their polarities will cancel and they will be nonpolar.

87 Which of the following is not correct for the ionic compound?

(A) Cation is form from losing the electron(s)
(B) Generally ionization energy of metal is lower than that of nonmetal
(C) Generally electron affinity of nonmetal is higher than that of metal
(D) Anion is formed from oxidation of neutral atom
(E) Ionic compound is formed from electrostatic force between cation and anion

88 The energy related with the formation of ionic solid between gaseous cation and anion is

(A) Bond energy   (B) Lattice energy

(C) Ionization energy   (D) Electron affinity

(E) Heat of vaporization

89 Which of the following is not correct for the chemical bond?

(A) There are covalent bond and ionic bond in $NH_4Cl(s)$

(B) There is ionic bond in $NaF(s)$

(C) The lattice energy of $MgCl_2$ is smaller than that of Al_2O_3

(D) Ionic compound can conduct electricity when it dissolved in water but can not conduct electricity when it is solid or liquid phase

(E) The melting point of $NaCl$ is lower than that of CaO

90 Which of the following is correct for the chemical bond?

I. The bond energy of $C-C$ bond is larger than that of $C=C$ bond

II. Generally the boiling point of ionic compound increases as lattice energy increases

III. Lattice energy can by calculated by Coulomb's law

(A) I only   (B) II only

(C) III only   (D) I, II only

(E) II, III only

91 Which of the following is correct for the geometry and polarity of molecule?

	Molecule	Molecular geometry	Polarity
(A)	PF_3	trigonal planar	polar
(B)	CF_4	square planar	nonpolar
(C)	$CHCl_3$	tetrahedral	nonpolar
(D)	OF_2	bent(V-shape)	polar
(E)	HF	linear	nonpolar

92 Which of the following is not correct for the molecule?

(A) Hybrid orbital of C in diamond is sp^3 but in graphite is sp^2

(B) Hybrid orbital of N in NH_3 is sp^2

(C) Hybrid orbital of C in CH_2CH_2 is sp^2 and geometry is trigonal planar

(D) The geometry of sp^3 can be tetrahedron

(E) π bond is the bond between p orbitals

93 Which of the following is not correct for the Lewis structure of CO_3^{2-}

(A) The Lewis structure of CO_3^{2-} has resonance
(B) The summation of oxidation number in CO_3^{2-} are -2
(C) The hybrid orbital of C is sp^2
(D) There is no pi(π) bond in CO_3^{2-}
(E) Central atom is satisfied with octet rule

94 Which of the following is correct for the VSEPR and VBT?

(A) BF_3 is T shape and hybrid orbital of B is sp^2
(B) H_2O is linear and hybrid orbital of O is sp
(C) The hybrid orbital of central atom in SO_2 and SO_3 are different
(D) The molecular geometry of C_2H_4 is trigonal planar and the hybrid orbital of C is sp^2
(E) The bond angle of CH_4 is 109.5° and the molecular geometry is square planar

95 Which substance has a polar covalent bond between its atoms?

(A) K_3N
(B) Ca_3N_2
(C) Na_2O
(D) F_2
(E) NH_3

96 Which of the following is correct for the chemical bond(s) found in $NaNO_3(s)$?

(A) Ionic bonds only
(B) Nonpolar covalent bonds only
(C) Ionic bond and polar covalent bond
(D) Polar covalent bond and dipole-dipole force
(E) Polar covalent bond only

97 Which of the following is not correct for the covalent bond in concerned with hybrid orbital and molecular geometry?

(A) The bond between p orbital and p orbital is always pi bond
(B) There are 7 sigma bond and 1 pi bond in CH_3COOH
(C) There are 3 sigma bond and 2 pi bond in C_2H_2
(D) The hybrid orbital of two oxygen atom in CH_3COOH are different
(E) Not all atoms of NH_3 are in the same plane

98 Which of the following is not correct for the molecular geometry, angles and hybrid orbital of central atom?

	Molecule	Molecular geometry	Bond angle	Hybrid orbital
(A)	H_2O	bent	smaller than $109.5°$	sp^3
(B)	BF_3	trigonal planar	$120°$	ssp^2
(C)	NH_3	trigonal pyramidal	smaller than $109.5°$	ssp^3
(D)	$BeCl_2$	linear	$180°$	ssp
(E)	O_3	bent	$90°$	ssp^3

99 Which of the following does not have the coordinate covalent bond?

(A) H_3O^+ (B) NO_2 (C) CO (D) SO_3 (E) CF_4

100 Which of the following can act as Lewis base?

(A) CH_4 (B) BF_3 (C) $BeCl_2$ (D) NH_3 (E) NH_4^+

Gas

6.1   Ideal Gas Law

6.1.1   Pressure of gas

- Barometer is used to measure the atmospheric pressure

  $1atm = 760mmHg = 760torr$

- Manometer is used to measure the pressure of specific gas

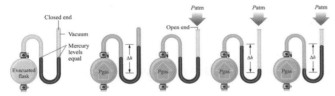

6.1.2   Ideal gas and real gas

Ideal gas	Characteristics	Real gas
○	mass	○
×	volume of gas particle	○
no repulsive force		repulsive force
×	intermolecular force	○
no phase change		phase change

6.1.3   Ideal gas law

- **Pressure-Volume relationship(constant n, and constant T) : Boyle's Law**

  The volume of gas is inversely proportional to pressure at constant temperature of given number of gas

  $P_1 V_1 = P_2 V_2$ from $PV = nRT$, $PV =$ constant

- **Temperature-Volume relationship(constant n, and constant P) : Charles's Law**

  The volume of gas is proportional to Kelvin temperature at constant pressure of given number of gas

$$\frac{V_1}{T_1} = \frac{V_2}{T_2} \quad \text{from } PV = nRT, \ \frac{V}{T} = \text{constant}$$

- **Temperature-Pressure relationship(constant n, and constant volume) : Gay-Lussac's Law**

  The pressure of gas is proportional to Kelvin temperature at constant volume of given number of gas

- **Temperature-volume-Pressure relationship(constant n) : The combined gas Law**

$$\frac{P_1 V_1}{T_1} = \frac{P_2 V_2}{T_2} \quad \text{from } PV = nRT, \ \frac{PV}{T} = \text{constant}$$

- **Density of gas**

STP(0℃, 1atm)	Non-STP
$\dfrac{M}{22.4L}$ (M : molar mass, g)	$\dfrac{PM}{RT}$

6.1.4 Dalton's law of partial pressure and water displacement

The total pressure of mixture of gas is the sum of the partial pressure of each component gas

$$P_T = P_A + P_B + P_C + \dots \quad (P_A, P_B, P_C \dots : \text{partial pressure})$$

$$\frac{P_A}{P_T} = \frac{n_A}{n_T} = \chi_A \ (\text{n : mol number}, \ \chi : \text{mole fraction})$$

In water displacement, the sum of the partial pressure of the gas and water vapor inside the collecting bottle is the same as the external pressure.

6.2 Kinetic Molecular Theory

6.2.1 Average Kinetic Energy and root mean square speed

- Average kinetic energy($\overline{E_k}$) of the molecule is proportional to Kelvin temperature regardless of kind of gas

$$\overline{E_k} = \frac{3}{2} kT \quad \overline{E_k} = \frac{3}{2} kT = \frac{1}{2} m v_{\text{rms}}^2 \quad v_{\text{rms}} = \sqrt{\frac{3kT}{m}} \ (v_{\text{rms}} : \text{root mean square speed})$$

6.2.2 Graham's Law

- Diffusion is the movement of gas particle from high concentration to low concentration
- Effusion is the movement of gas particle through pin hole into evacuated container

$$\frac{v_B}{v_A} = \sqrt{\frac{m_A}{m_B}} \quad \text{at same temperature and pressure}$$

Chapter Review Questions 6. Gas

[101-102] **Refer to the following**

(A) Ideal gas constant
(B) Celsius temperature
(C) Kelvin temperature
(D) Partial pressure
(E) Volume

---

101 is proportional to moles of gas, when pressure and temperature are held constant

102 is proportional to the mole of gas of mixture in the container

103 The average speed of hydrogen gas is twice as that of oxygen gas at constant temperature

Because

Average kinetic energy of gas is proportional to only kelvin temperature regardless of type of gas

104 The pressure of gas in rigid container proportional to mole number of gas at constant temperature.

Because

The average kinetic energy is proportional to the kelvin temperature.

105 The average speed of lighter gas molecule is larger than that of heavier gas molecule at the same temperature

Because

Heavier gas and lighter gas have the same kinetic energy at same temperature

106 An ideal gas differs from a real gas in that ideal gas does not experience intermolecular force

Because

The collision of ideal gas is elastic but that of real gas is not elastic

107 When a sample of gas is heated at constant pressure from 300K to 600K, final volume is 2 times larger than its original volume

Because

The volume of each gas particle is doubled

108 Which of the following conditions the actual gas exhibits the closest behavior to the ideal gas?

I. High temperature
II. Low pressure
III. Low molecular weight

(A) I only   (B) II only   (C) III only   (D) I, II only   (E) I, II, and III

109 There are same number of Ne and H_2 gas in rigid 3.0L of container. Which of the following is (are) correct?

I. Average kinetic energy of each gas is the same
II. Average speed of each gas is the same
III. Partial pressure of each gas is the same

(A) I only   (B) II only   (C) III only   (D) I and II only   (E) I and III

110 There are two 1L of vessels. Each vessel contains different gas sample. The followings are the condition of each gas

	Gas A	Gas B
Pressure	1atm	2atm
Temperature	400K	500K

Which of the following is the mole ratio of gas A compare to gas B?

(A) 1 : 2   (B) 2 : 1   (C) 4 : 5   (D) 5 : 8   (E) 2 : 5

111 Consider three 1L flasks at STP. Flask A contains CO_2 gas, flask B contains CH_4, and flask C contains O_2 gas. Which flask contains the largest number of molecule?

(A) Flask A
(B) Flask B
(C) Flask C
(D) All flasks contain same number of molecules because the number of gas molecule is always constant regardless temperature and pressure
(E) All flasks contains same number of molecules because they are the same temperature and pressure

112 5.0atm He gas in 2.0L container and 2.5atm Ar gas in 4.0L container are mixed completely 5.0L of container.
Which of the followings is correct of partial pressure for each gas?

	He	Ar			He	Ar
(A)	5.0atm	2.5atm		(B)	2.5atm	5.0atm
(C)	2.0atm	2.0atm		(D)	2.0atm	4.0atm
(E)	4.0atm	2.0atm				

113 If 3.0mol of He gas and 2.0mol of Ar gas is mixed completely in rigid vessel. The total pressure is 6.0atm. Which of the following is correct for the partial pressure of each gas?

	He	Ar			He	Ar
(A)	6.0atm	6.0atm		(B)	3.6atm	2.4atm
(C)	1.8atm	4.2atm		(D)	3.0atm	3.0atm
(E)	1.5atm	4.5atm				

114 Which of the following gases is most dense when all are measured under the same temperature and pressure?(The molar mass of H : 1g/mol, C : 12g/mol, N : 14g/mol, O : 16g/mol, F, 19g/mol)

(A) CH_4 (B) N_2 (C) O_2 (D) F_2 (E) CO_2

115 The density of gas is 2.05g/L. at STP. Which of the following gas is expected to be this gas?(The molar mass of H : 1g/mol, C : 12g/mol, N : 14g/mol, O : 16g/mol, S : 32g/mol)

(A) NH_3 (B) CH_4 (C) CO_2 (D) NO_2 (E) SO_2

116 Which of the following gases can be collected through water displacement?

I. H_2
II. CO_2
III. NO_2

(A) I only   (B) II only   (C) III only   (D) I and II only   (E) I, II and III

117 When adding the He(g) in rigid container containing Ar(g). Which of the following is not correct? (There is no reaction between gas molecules and finally, the mole fraction of two gases are the same)

(A) The average kinetic energy of each gas is the same
(B) The temperature of container is decrease
(C) The density of gas mixture is increase
(D) The pressure of gas mixture is increase
(E) It the container is opened helium gas is diffused faster than argon

[118-119] **The below is the unbalanced decomposition reaction of potassium chlorate**

$$...KClO_3(s) \rightarrow ...KCl(s) + ...O_2(g)$$

118 The oxygen produced was collected by water displacement at 25℃ and pressure of 760mmHg. Which of the following is correct for the partial pressure of $O_2(g)$ in the gas collected?(The vapor of water is 24mmHg at 25℃)

(A) 736mmHg   (B) 760mmHg   (C) 784mmHg   (D) 808mmHg   (E) 832mmHg

119 If 122g of $KClO_3$ were completely decomposed, what mass of KCl(s) would be expected to be produced?(at 25℃, 1atm, the volume of 1mol of gas is 24.5L, the molar mass of $KClO_3$: 122g/mol, O_2 : 32g/mol)

(A) 56g   (B) 74g   (C) 92g   (D) 104g   (E) 122g

120 The mass of 3.01×10^{23} gas particles at STP is 14g. Which of the following is not correct?
(A) The density of gas is 1.25g/L at STP
(B) The molar mass is 28g/mol
(C) The density increase as mole number increase
(D) The average speed of this gas is larger than that of oxygen gas at the same temperature
(E) The average kinetic energy of this gas is the same as that of hydrogen gas at same temperature

07   Solid, Liquid, and Phase Change

7.1   Properties of Solid, Liquid, and Gas

Properties	Solid	Liquid	Gas
Volume	definite volume	definite volume	the same as container
Shape	definite shape	the same as container	the same as container
Density	high	high	low
Compressibility	incompressible	a little compressible	very compressible
Motion and movement	vibration in fixed position	relatively move freely	very freely move

7.2   Crystalline Structure

7.2.1   Types of Crystalline structure

Crystalline is the structure with long-range order of its particles.

• **Atomic crystalline(network solid)**

Atom are joined constituent network covalent bond. Though these structures are formed through covalent bond, they do not form a molecules. **ex** diamond(C), graphite(C), quartz(SiO_2),...

• **Metallic crystalline**

Metallic crystalline is composed through the electrostatic attraction between cation and free electron(s). The characteristics of metallic crystalline, such as electricity, ductility and malleability are due to the free electron(s) **ex** Na, Mg, Al, Fe,...

• **Ionic crystalline**

Ionic crystalline is formed from the electrostatic attraction between cation(s) and anion(s)
ex NaCl, MgO, CaF_2, Fe_2O_3,...

- **Molecular crystalline**

Molecular crystalline is formed through intermolecular force. The melting and boiling points are lower than other crystalline structure because intermolecular force is not the chemical bond

Crystalline	Composition	Particles of crystalline	Types of bond	Melting and Boiling point	Electricity
Atomic crystalline	Nonmetal	Atom	Covalent bond	Very high	No but Graphite (graphene)
Metallic crystalline	Metal	Cation-Electron	Metallic bond	High	Yes
Ionic crystalline	Metal-Nonmetal	Cation-Anion	Ionic bond	High	Yes in Molten or aqueous solution
Molecular crystalline	Nonmetal	Molecule	Intermolecular force	Low	No

7.2.2 Crystalline structure and amorphous structure

- Crystalline structures have constant value of characteristics of matter, such as bond energy, density, melting and boiling point
- Amorphous structures, such as glass(SiO_2), lack the long range order of particles and do not form the lattice structure. The bond energy, density, melting and boiling point are not constant

7.3   Intermolecular Force

7.3.1 Intermolecular force

Intermolecular force are attractive force between molecules

- **Dipole − dipole force**

Dipole − dipole force is attractive force between polar molecules with permanent dipole

- **Instantaneous dipole − induced dipole force(London dispersion force; LDF)**

Instantaneous dipole is formed from the polarized of electrons in molecule and instantaneous dipole induce the temporary dipole of other adjacent molecules or atoms, All molecules have LDF.

- **Hydrogen bond**

Hydrogen bond is another type of dipole − dipole force between the molecules with H − N, H − O, H − F as chemical bonds. This is very strong intermolecular force because H is highly electropositive and N, O, F are highly electronegative

7.3.2 Characteristics of water

Characteristics of water are due to the strong intermolecular force, hydrogen bond

• **Water as solvent**

Polar molecules and may of ionic compounds is dissolved in watere. The process in which a ion or a molecule is surrounded by water molecule is hydration

• **Surface tension**

Cohesive force	Adhesive force
Intermolecular attraction between like molecules	Intermolecular attraction between unlike molecules

Surface tension is the energy needed to increase the surface area of liquid. Liquids with strong intermolecular force have large surface tension. capillary action is the phenomenon in which a liquid rises or falls along a thin tube and is related to surface tension.

• **Viscosity**

Viscosity formed form intermolecular force is the resistance of fluid . As the viscosity of matter increases, the fluidity decreases

• **Density of water**

Generally the density of matter increases in order gas, liquid, solid. But the density of water is larger than that of ice When water freezes to ice, the volume increases. Because space is formed inside the molecular crystal by hydrogen bond

7.4   Phase Change

7.4.1 Phase change

• **Vaporization and condensation**

At a certain temperature, the liquid molecules with sufficient kinetic energy escape from the surface. this is the vaporization. Some molecules in vapor phase return to the liquid phase. This is the condensation

• **Melting and freezing**

Melting is the phase change process from solid to liquid. Freezing is the phase change process from liquid to solid

• **Sublimation and deposition**

Sublimation is the phase change process from solid to gas. Deposition is the phase change process from gas to solid

7.4.2 Vapor pressure

- Vapor pressure is the pressure exerted by a vapor when the vaporization and condensation rate reach to equilibrium
- Vapor pressure is dependent on temperature. As increasing the temperature, the vapor pressure is increasing
- The boiling point is the temperature of liquid when the vapor pressure is equal to the pressure above its surface Particularly, this pressure is 1atm(760mmHg), the temperature is referred to as the normal boiling point.

Intermolecular force ↑	→	Vapor pressure ↓	→	Boiling point ↑	→	Heat of vaporization ↑
		At same temperature		At same pressure		

7.4.3 Phase diagram

Phase diagram is the graph showing which phases are the stable at certain temperature and pressure condition

- **Triple point and critical point**

  Triple point is the point at which all three phases are in equilibrium. Critical point is composed of critical temperature and critical pressure. When pressure is added to gas at above the critical temperature, the gas and liquid phases are not distinguishable and this is called supercritical fluid

- **Phase diagram of CO_2**

  The density is increasing in order gas, liquid, solid. Liquid phase can not be observed at 1atm because the pressure of triple point is higher than 1atm

- **Phase diagram of H_2O**

  The density is increasing in order gas, solid, liquid. When pressure is added to solid, the phase change from solid to gas.

Chapter Review Questions ▶ **7. Solid, Liquid, and Phase Change**

[121-122] Refer to the following

(A) Deposition
(B) Sublimation
(C) Liquefaction
(D) Fusion
(E) Vaporization

---

121 is the process of phase change from liquid to gas

122 is main process of phase change of dry ice in room temperature

---

[123-124] Refer to the following

(A) Na
(B) C(graphite)
(C) CO_2
(D) NaCl
(E) H_2O

---

123 is network solid with delocalized electrons but not metal

124 is the substance with the lowest intermolecular force

125 The temperature of a substance always increase as heat energy is added to the system

Because

The average kinetic energy of the particles in the system increase with an increase in temperature

126 For H_2O, liquid, unlike most substances, is denser than solid

Because

Water molecules can make hydrogen bonds with neighboring molecules.

127 The boiling point of H_2O is higher than that of HF

Because

The average number of hydrogen bond in a H_2O molecules is larger than HF

128 Which of the following is correct ?

I. Boiling point of H_2O is relatively higher than that of other nonpolar molecules with similar molar mass because of strong covalent bond between H and O atom in H_2O molecule

II. Boiling point of nonpolar molecule is always lower than that of polar molecule

III. Boiling point of F_2 is lower than that of Cl_2 because the molar mass of F_2 is smaller than that of Cl_2

(A) I only   (B) II only   (C) III only   (D) I, II only   (E) II, III only

129 Which of the following is not correct for the ionic compound?

(A) Ionic compounds are formed by the electrostatic attraction of cation(s) and anion(s).
(B) Generally, ionic compounds have a higher melting point than molecules
(C) Ionic compounds can flow current in both molten and aqueous solution
(D) The high processability of ionic compounds is due to the strong ionic bonding.
(E) In ionic compounds, cations and anions are bonded continuously and there is no molecule

130 Which of the following is correct when the equilibrium state is reached after placing the water in a sealed vacuum vessel at 20℃ ?

(A) The rate of vaporization is the same as that of condensation and reaches liquid–gas equilibrium
(B) The rate of vaporization is the same as that of deposition and reaches solid–gas equilibrium.
(C) The mole number of liquid phase and gas phase are the same
(D) Reaches the triple point
(E) Reaches the critical point

131 Which of the following is the definition of boiling point?

(A) The temperature when liquid is vaporized in gas phase
(B) The temperature when the volume of liquid is decreased
(C) The temperature when intermolecular force between liquid molecules is increased
(D) The temperature when vapor pressure is the same as atmospheric pressure or external pressure
(E) The temperature when vapor pressure is the same as 760mmHg

132 Which of the following is the correct order in increasing the boiling point?

(A) HF < HCl < HBr < HI   (B) HCl < HF < HI < HBr
(C) HI < HBr < HCl < HF   (D) HCl < HBr < HI < HF
(E) HBr < HI < HCl < HF

133 Which of the following is correct for the bellow graph for the vapor pressure?

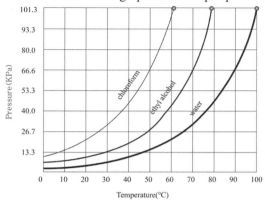

(A) Intermolecular force between water(H_2O) molecule is larger than that of ethanol(C_2H_5OH) because molecular mass of water is larger than that of ethanol

(B) Intermolecular force between water molecules is only hydrogen bond

(C) Chloroform is not vaporized below $60°C$

(D) Ethanol is not dissolved in water molecule because ethanol in nonpolar molecule

(E) Each line of above graph represents the equilibrium state between liquid and gas phases

134 Which of the following is correct explanation for the reason that boiling point of I_2(iodine) is higher than that of hydrogen fluoride(HF)

(A) I_2 is polar molecule but HF is nonpolar molecule

(B) Molar mass of I_2 is larger than that of HF

(C) Intermolecular force between I_2 molecules is larger than HF because HF molecules have only London dispersion force

(D) HF molecule has polar covalent bond but I_2 molecule has nonpolar covalent bond

(E) More electrons can be polarizable in HF molecules than I_2

135 Carbon dioxide can sublimated at room temperature and 1atm. Which of the following is the correct explanation for this observation?

(A) The pressure of triple point is lower than 1atm

(B) The pressure of triple point is higher than 1atm

(C) The density of liquid carbon dioxide is smaller than solid carbon dioxide

(D) Liquid carbon dioxide can not be exist at any condition

(E) The pressure of critical temperature of carbon dioxide is too high to be liquefied

136 Which of the following is correct for the intermolecular force?

(A) The abnormally high boiling point of water in comparison to compounds with similar molar mass is due primarily to LDF

(B) The intermolecular force in polar molecule is only dipole-dipole force

(C) The soluble ionic compound forms ion-dipole attraction when it dissolves in water

(D) There are more hydrogen bonds in methane(CH_4) molecules than water(H_2O) molecules

(E) Both network solid and molecular crystalline structure are formed through chemical bond

137 Which of the following is not correct for the intermolecular force in each molecule

(A) H_2O London dispersion force, dipole − dipole force, hydrogen bond

(B) CH_3F London dispersion force, dipole − dipole force, hydrogen bond

(C) C_2H_6 London dispersion force

(D) SCl_2 London dispersion force, dipole − dipole force

(E) CO_2 London dispersion force

138 Which of the following is not for correct for the chemical bond?

(A) CH_3CH_2OH is formed form covalent bond between each atom

(B) NaCl(ionic crystalline)is formed form ionic bond between Na^+ ion and Cl^-ion

(C) C(graphite) and C(diamond) is formed network covalent bond and they are isotope

(D) SiO_2 is formed network covalent bond and shows high melting and boiling point

(E) Amorphous substance, such as glass, lacks a well defined arrangement and long-range molecular order

139 Which of the following is not correct for the chemical bond?

(A) The high melting and boiling point of diamond are from network covalent bond

(B) Metallic bond is related with free electrons(valence electrons) and cation

(C) The melting point of Na is higher than that of K

(D) Both diamond and graphite have electroconductivity because of their free electron

(E) Generally the boiling point of ionic compound increase as lattice energy increase

140 The below is the vapor pressure of some liquids at 20℃

Liquid	Vapor pressure(20℃)
A	20
B	35
C	40
D	55
E	100

Which of the following has the lowest boiling point at the same atmospheric pressure?

(A) A   (B) B   (C) C   (D) D   (E) E

08 Solution

8.1 Solution

8.1.1 Classification of Mixture

Type	Size(nm)	Examples	Properties
Solution	~2.0	Air, Sea water	Transparent to light, Precipitation is not occurred
Colloid	2.0~100	Milk, Fog, Butter	Opaque to light, Precipitation is not occurred and filtered
Suspension	100~	Blood, Spray, Aerosol	Opaque to light, Precipitation is occurred and filtered

8.1.2 Colloid

Properties of Colloid	
Brown motion	**Tyndall effect**
The random movement of colloidal particles suspended in a liquid or gas	The scattering of light by particles
Coagulation	

① Colloidal particles show the charge in medium : electrostatic repulsion
② Coagulation can be accomplished either by heating or by adding an electrolyte
  : Heating → increase the velocity of the colloidal particles and aggregate
  : Adding an electrolyte → neutralize the charge of colloidal particle

8.1.3 Solution : Homogeneous mixture

• **Solution**

  The homogeneous mixture composed of solvent and solute

• **Solvent**

  The dissolving medium in solution

• **Solute**

  The substance dissolved in a solvent

Solute	Solvent	State of solution	Example
Gas	Gas	Gas	air : $N_2(g) + O_2(g)$,...
Gas	Liquid	Liquid	soda : $H_2O(l) + CO_2(g) \rightarrow H_2CO_3(aq)$
Gas	Solid	Solid	$H_2(g)$ in $Pd(s)$ (palladium)
Liquid	Liquid	Liquid	$H_2O(l) + CH_3CH_2OH(l) \rightarrow CH_3CH_2OH(aq)$
Solid	Liquid	Liquid	$H_2O(l) + NaCl(s) \rightarrow NaCl(aq)$
Solid	Solid	Solid	brass : $Cu(s) + Zn(s)$, solder : $Sn(s) + Pb(s)$

8.2　Concentration

8.2.1　Expression of concentration

% concentration	Molarity	Molality	Mole fraction
$\dfrac{\text{weight of solute}}{\text{weight of solution}} \times 100$	$\dfrac{\text{mole of solute(mol)}}{\text{volume of solution(L)}}$	$\dfrac{\text{mole of solute(mol)}}{\text{mass of solvent(kg)}}$	$\dfrac{\text{mol of component}}{\text{total mol number}}$

8.2.2　Solubility

Solubility is the maximum amount of solute dissolved in given amount of solvent at a certain temperature. Generally solubility is expressed as g/100g solvent

Saturated solution	contains the maximum amount of solute dissolved
Unsaturated solution	can still dissolve more solute
Supersaturated solution	contain more solute than it theoretically can; crystal nucleus will make it deposit

• **Solubility of solid solute in water**

As the temperature increases, the solubility increases because generally, solid solute dissolving process is endothermic. The solubility is not dependent on pressure because compressibility of solid is nearly zero

• **Solubility of gas solute in water**

As the temperature increases, the solubility decrease because gas solute dissolving process is exothermic. The solubility can be increased by increasing the partial pressure for the gas with low solubility in water

Henry' Law　$C_g = k_H P_g$

C_g : solubility(M)　K_H : Henry's constant(M/atm)　P : partial pressure of gas(atm)

8.2.3 Making the Standard Solution

• **Standard solution**

A solution of accurately known concentration. The standard solution is used to know another solution of unknown concentration through stoichiometric relationship in chemical reaction

• **Preparation of standard solution**

1) Making the solution through dissolving the solid solute to the solvent

   ⓔⓧ 3.00M of 100.mL NaOH(aq) preparation

   12g of NaOH(s) is added to a volumetric flask with distilled water less than final volume of solution and distilled water is added to the 100mL mark

2) Making the solution through dilution of concentrated solution

   ⓔⓧ 3.00M of 100.mL NaOH(aq) preparation from 5M of stock solution

   Aliquot 60 mL of 5M aqueous solution and transfer it to the flask, and then add the distilled water to 100L mark

8.3   Chemical Reaction of Solution

8.3.1 Electrolyte and Nonelectrolyte

• **Electrolyte**

When electrolyte is dissolved in water, the solution conducts electricity (strong and weak electrolyte)

• **Nonelectrolyte**

Nonelectrolyte does not conduct electricity when it is dissolved in water

8.3.2 Chemical reaction of aqueous solution

• **Activity series of metal and single replacement reaction**

 Activity series of metal

K > Ca > Na > Mg > Al > Zn > Fe > Ni > Sn > Pb > (H) > Cu> Hg > Ag > Pt > Au

← chemical reactivity increase   chemical reactivity decrease →
relatively easily oxidized   relatively hardly oxidized
good reducing agent   poor reducing agent

$Cu(NO_3)_2(aq) + Zn(s) \rightarrow Zn(NO_3)_2(aq) + Cu(s)$　　　$Zn(NO_3)_2(aq) + Cu(s) \rightarrow$ No reaction

$Na(s) + 2H_2O(l) \rightarrow 2NaOH(aq) + H_2(g)$　　　$Zn(s) + 2H_2O(l) \rightarrow$ No reaction

$2Na(s) + 2HCl(aq) \rightarrow 2NaCl(g) + H_2(g)$　　　$Zn(s) + 2HCl(aq) \rightarrow ZnCl_2(g) + H_2(g)$

• **Double replacement reaction and precipitation**

$NaCl(aq) + AgNO_3(aq) \rightarrow NaNO_3(aq) + AgCl(s)$　　　$Ca(NO_3)_2(aq) + Na_2SO_4(aq) \rightarrow 2NaNO_3(aq) + CaSO_4(s)$

$NH_4NO_3(aq) + NaCl(aq) \rightarrow NH_4Cl(aq) + NaNO_3(aq)$　　$Pb(NO_3)_2(aq) + 2NaI(aq) \rightarrow 2NaNO_3(aq) + PbI_2(s)$

• **Redox reaction**

Combustion reaction, synthesis and decomposition reaction, single replacement reaction are oxidation−reduction(redox) reaction

combustion reaction	$C_3H_8(g) + 5O_2(g) \rightarrow 3CO_2(g) + 4H_2O(l)$
synthesis reaction	$3Mg(s) + N_2(g) \rightarrow Mg_3N_2(s)$
decomposition reaction	$2HgO(s) \rightarrow 2Hg(l) + O_2(g)$
single replacement reaction	$2Fe(s) + 3Cu(NO_3)_2(aq) \rightarrow 2Fe(NO_3)_3(aq) + 3Cu(s)$

• **Net ionic equation**

Net ionic equation is the chemical equation expressed by the ions which are involved in reaction. spectator ion is the ion unchanged in reaction

$2NaI(aq) + Pb(NO_3)_2(aq) \rightarrow 2NaNO_3(aq) + PbI_2(s)$　　　　　　　molecular equation

$2Na^+(aq) + 2I^-(aq) + Pb^{2+}(aq) + 2NO_3^-(aq) \rightarrow 2Na^+(aq) + 2NO_3^-(aq) + PbI_2(s)$ complete ionic equation

$2I^-(aq) + Pb^{2+}(aq) \rightarrow PbI_2(s)$　　　　　　　　　　　　　　　net ionic equation

net ion : $Pb^{2+}(aq)$, $I^-(aq)$　　　spectator ion : $Na^+(aq)$, $NO_3^-(aq)$

• **Solubility rule**

Soluble compounds	Exception
Contain ion of alkali metal Li^+, Na^+, K^+,...	
NO_3^-, NH_4^+, $C_2H_3O_2^-$,...	
Cl^-, Br^-, I^-	Ag^+, Hg_2^{2+}, Pb^{2+}
SO_4^{2-}	Ag^+, Ca^{2+}, Ba^{2+}, Hg_2^{2+}, Pb^{2+}

Insoluble compounds	Exception
CO_3^{2-}, PO_4^{3-}, S^{2-}	Compounds containing alkali metal ions
$OH-$	Compounds containing alkali metal ions, Ca^{2+}, Ba^{2+}

Precipitate	Color
$AgCl(s)$	White
$AgBr(s)$	Bright Yellow
$AgI(s)$	Yellow
$PbI_2(s)$	Yellow
$PbCrO_4(s)$	Yellow
$CuO(s)$	Black
$CdS(s)$	Yellow
$PbS(s)$	Black
$Al(OH)_3$	White
$BaCO_3(s)$	White
$BaSO_4(s)$	White
$CaCO_3(s)$	White
$CaSO_4(s)$	White

• **Neutralization Reaction**

HNO₃(aq) + NaOH(aq) → H₂O(l) + NaNO₃(aq)	
Net ionic equation of neutralization	$H^+(aq) + OH^-(aq) \rightarrow H_2O(l)$
Net ionic equation	$H^+(aq) + OH^-(aq) \rightarrow H_2O(l)$
H₂SO₄(aq) + Ba(OH)₂(aq)→ 2H₂O(l) + BaSO₄(s)	
Net ionic equation of neutralization	$H^+(aq) + OH^-(aq) \rightarrow H_2O(l)$
Net ionic equation	$2H^+(aq) + SO_4^{2-}(aq) + Ba^{2+}(aq) + 2OH^-(aq) \rightarrow 2H_2O(l) + BaSO_4(s)$
HC₂H₃O₂(aq) + NaOH(aq) → H₂O(l) + NaC₂H₃O₂(aq)	
Net ionic equation of neutralization	$HC_2H_3O_2(aq) + OH^-(aq) \rightarrow H_2O(l) + C_2H_3O_2^-(aq)$
Net ionic equation	$HC_2H_3O_2(aq) + OH^-(aq) \rightarrow H_2O(l) + C_2H_3O_2^-(aq)$
HCl(aq) + NH₃(aq) → NH₄Cl(aq)	
Net ionic equation of neutralization	$H^+(aq) + NH_3(aq) \rightarrow NH_4^+(aq)$
Net ionic equation	$H^+(aq) + NH_3(aq) \rightarrow NH_4^+(aq)$

8.4   Colligative properties of Solution

Colligative properties is the characteristics of solution depend on the number of solutes in solution regardless of nature of solutes

8.4.1   Vapor pressure(Raoult's Law)

• **Non-volatile Nonelectrolyte**

When the volatile solute B is dissolved in solvent A

$$P_A = P_A^{\circ}x_A$$

P_A : vapor pressure of solution x_A : mole fraction of A

vapor pressure depression$(\Delta P) = P_A^{\circ} - P_A^{\circ}x_A = P_A^{\circ}(1 - x_A) = P_A^{\circ}x_B$ $(x_A + x_B = 1)$

• **Volatile Nonelectrolyte**

When the volatile solute B is dissolved in solvent A

$$P_{sol} = P_A + P_B = P_A^{\circ}x_A + P_B^{\circ}x_B(x_A + x_B = 1)$$

$P_A = P_A^{\circ}x_A$: vapor pressure of A in solution $P_B = P_B^{\circ}x_B$: vapor pressure of B in solution

8.4.2 Boiling point elevation and freezing point depression

For nonvolatile nonelectrolyte solute

- Boiling point elevation(℃, or K)　$\Delta T = k_b m$
  k_b : boiling point elevation constant (℃/m)　m : molality(m)
- Freezing point lowering(℃, or K)　$\Delta T = k_f m$
  k_f : freezing point lowering constant (℃/m)　m : molality(m)

8.4.3 Osmotic pressure

Osmosis is movement of solvent from lower concentration to higher concentration through semi-permeable membrane

$$\pi = cRT$$
π : osmotic pressure(atm)　c : molarity(M)　R : gas constant(atm · L/mol · K)　T : Kelvin temperature(K)

8.4.4 van't Hoff factor(i) in colligative properties

van't Hoff factor is considered in colligative properties of electrolyte solution

$$\Delta T = k_b m i \qquad \Delta T = k_f m i \qquad \pi = cRT i$$
(i : theoretically for $C_6H_{12}O_6$ is 1, NaCl is 2, $Ca(NO_3)_2$ is 3,...)

※ from the colligative properties, molar mass of solute can be measured

Chapter Review Questions | **8. Solution**

[141-143] **Refer to the following**

(A) combustion

(B) decomposition

(C) synthesis

(D) single replacement reaction

(E) double replacement reaction

141 The reaction between $KI(aq)$ and $AgNO_3(aq)$

142 The reaction of hydrocarbon can produce $CO_2(g)$ and $H_2O(l)$

143 The reaction between $Zn(s)$ and $CuSO_4(aq)$

144 $HCl(aq)$ is a strong electrolyte and $HC_2H_3O_2(aq)$ is a weak electrolyte.

Because

$HCl(aq)$ is 100% ionized but $HC_2H_3O_2(aq)$ is partially ionized in water.

145 In general, the solubility of solid materials increases with increasing temperature

Because

The dissolution process of the solid solute is generally an endothermic reaction

146 The freezing point of 0.2m sugar solution is lower than that of 0.2m of potassium chloride solution

Because

potassium chloride is ionized in the aqueous solution while sugar does not

147 At constant pressure, the boiling point of the aqueous solution is higher than the pure water

Because

The vapor pressure of the aqueous solution is higher than the pure water

148 Which of the following is correct for the reaction that 2.0L of 1.0M $AgNO_3$(aq) and 3.0L of 1.0M of NaCl(aq) are mixed?

I.   1mol of white precipitate is formed
II. Net ionic equation is Ag^+(aq) + Cl^-(aq) → AgCl(s)
III. The concentration of spectator ions in the mixture is the same.

(A) I only   (B) II only   (C) I and II only   (D) II and III only   (E) I, II and III

149 Which of the following is correct?

I.   The color of $KMnO_4$(aq) solution is purple
II. When Zn(s) is added to $Cu(NO_3)_2$(aq), the color of solution change from blue to colorless
III. When $Pb(NO_3)_2$(aq) and NaI(aq) are mixed, the white precipitate is formed

(A) I only   (B) II only   (C) I and II only   (D) II and II only   (E) I, II and III

150 Which of the following is correct for the reaction represented below?

$$Cu(NO_3)_2(aq) + 6NH_3(aq) \rightarrow [Cu(NH_3)_6]^{2+}(aq) + 2NO_3^-(aq)$$

I. $Cu(NO)_3$(aq) is reducing agent
II   The color of solution does not change when the forward reaction proceeds
III. $[Cu(NH_3)_6]^{2+}$(aq) is the complex ion

(A) I only   (B) II only   (C) I and II only   (D) II and III only   (E) I, II and III

151 Which of the following is correct for the reaction between $HC_2H_3O_2$(aq) and $CaCO_3$(s)

I.   The reaction is neutralization reaction
II. White precipitate is formed
III. CO_2(g) gas is evolved

(A) I only   (B) II only   (C) I and II only   (D) I and III only   (E) II and III

152 Which of the following is correct for the osmosis and osmotic pressure?

I.   When the two compartment separated by semipermeable membrane, solvent molecules can move through the membrane but solute molecules can not
II. The osmotic pressure is proportional to the molarity and temperature of the solution.
III. The osmotic pressure of 0.1M of NaCl(aq) and 0.05M sugar solution at the same temperature is the same.

(A) I only   (B) II only   (C) I and II only   (D) II and II only   (E) I, II and III

153 Which of the following is correct for the definition of molarity?

(A) The molar mass of solute in gram per 1L of solvent
(B) The mol number of solute per 1L of solution
(C) The value required to change the concentration of 1mol/kg solution to the molar concentration
(D) The mol number of solute to total mol number of solution
(E) The mass of solute per 100g of solution

154 Which of the following is correct value of the concentration of Cl^-(aq) in aqueous solution by mixing 2.0L of 2.0M NaCl(aq) and 3.0L of 3.0M $CaCl_2$(aq)?

(A) 1.8M   (B) 2.6M   (C) 3.6M   (D) 4.4M   (E) 5.0M

155 What is the molar concentration of 2.5m NaOH(aq)? (molar mass of NaOH : 40g/mol, density of NaOH(aq) : 1.2g/mL)

(A) 2.5M   (B) 2.7M   (C) 4.0M   (D) 4.2M   (E) 5.0M

156 Which of the following has the highest electrical conductivity?

(A) 0.2M of $C_6H_{12}O_6$(aq)   (B) 0.2M of $HC_2H_3O_2$(aq)   (C) 0.3M of H_3PO_4(aq)
(D) 0.2M of NaOH(aq)   (E) 0.2M of C_2H_5OH(aq)

157 What is the volume of 10m concentration of NaOH(aq) required to make a 0.5L aqueous solution of 2.0M NaOH(aq)?(molar mass of NaOH : 40g/mol, density of NaOH(aq) solution : 1.4g/mL)

(A) 100mL   (B) 200mL   (C) 300mL   (D) 400mL   (E) 500mL

158 Which of the following is the mass of $Ca(OH)_2$ that should be added to 200 g of water to make a 0.4 m aqueous $Ca(OH)_2$ solution? (molar mass of $Ca(OH)_2$: 74g/mol)

(A) 2.96   (B) 5.92   (C) 7.40   (D) 11.8   (E) 14.8

159 Which of the following correct Except for the solution?

(A) Vapor pressure of solution is lower than that of pure solvent at same temperature
(B) Boiling point of solution is higher than that of pure solvent at same atmospheric pressure
(C) The boiling point 0.1m of $CaCl_2$(aq) is higher than that of 0.1m of NaCl(aq)
(D) Colligative properties is the characteristics of solution that depend on the number of solute particles
(E) Freezing point 0.1m of NaCl(aq) is lower than that of 0.1m of sucrose solution because solubility of NaCl is higher than that of sucrose at the same temperature

160 Which of the following is the correct boiling point when 40g of NaOH and 180 g of $C_6H_{12}O_6$ is each dissolved in 1kg of water(the molar mass of NaOH and $C_6H_{12}O_6$: 40g/mol, 180g/mol and atmospheric pressure : 1atm, molal boiling point elevation constant (k_b) : 0.5°C/m)

	NaOH(aq)	C6H12O6(aq)
(A)	100℃	100℃
(B)	100.5℃	100.5℃
(C)	101℃	100.5℃
(D)	101℃	101℃
(E)	100.5℃	101℃

09 Thermochemistry

9.1 Heat of Reaction

9.1.1 Exothermic and Endothermic Reaction

• Exothermic reaction

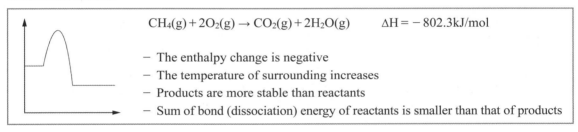

$$CH_4(g) + 2O_2(g) \rightarrow CO_2(g) + 2H_2O(g) \qquad \Delta H = -802.3 kJ/mol$$

 − The enthalpy change is negative
 − The temperature of surrounding increases
 − Products are more stable than reactants
 − Sum of bond (dissociation) energy of reactants is smaller than that of products

• Endothermic reaction

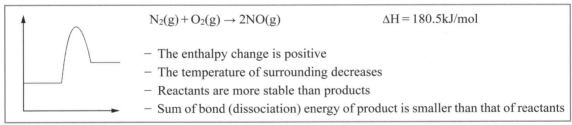

$$N_2(g) + O_2(g) \rightarrow 2NO(g) \qquad \Delta H = 180.5 kJ/mol$$

 − The enthalpy change is positive
 − The temperature of surrounding decreases
 − Reactants are more stable than products
 − Sum of bond (dissociation) energy of product is smaller than that of reactants

9.1.2 Calorimetry

• Specific heat (capacity) (c : cal/g℃, J/g℃)

Specific heat is the amount of heat required to raise the temperature of 1g of the substance by 1℃. The specific heat (capacity) is the characteristics of matter

• Heat capacity (C : cal/℃, J/℃)

Heat capacity is the amount of heat required to raise the temperature of given mass of the substance by 1℃. The heat capacity is not the characteristics of matter

- Calculation of heat of reaction(Q) from the calorimetry

$$Q(cal) = c \times m \times \Delta T$$

c : specific heat capacity (cal/g℃)   m : mass of sample(g) ΔT : temperature difference (T_{final}-$T_{initial}$)

- Standard enthalpy of formation

Standard enthalpy of formation is the heat change when 1mol of the compound is formed from the its most stable elements at standard state(25℃, 1atm) and the standard enthalpy of formation of the most stable element is 0

$$\Delta H = \Sigma n \Delta H_f^o (products) - \Sigma n \Delta H_f^o (reactants)$$

Standard enthalpy of reaction is the enthalpy of a reaction carried out at standard states.

- Bond (dissociation) energy (BE)

The bond (dissociation) energy is the enthalpy change when 1mol of gaseous covalent bond is broken at standard state(25℃, 1atm)

$$\Delta H = \Sigma BE(reactants) - \Sigma BE(products)$$

- Hess's Law

The enthalpy change is the same when the reactants and products are same whether the reaction takes place in one step or in a series of steps.

9.2   Thermodynamics

9.2.1 Thermodynamic system and data

- Isolated system

None of matter and energy can be transferred between system and surrounding in isolated system

- Closed system

Energy can be transferred but matter is not between system and surrounding in closed system

- Open system

Matter and energy(work, heat) can be transferred between system and surrounding in open system

- State function

State functions are properties that are determined by the state of the system regardless of the path for example, enthalpy(ΔH), entropy(S, ΔS), free energy(ΔG), internal energy(E, ΔE)

• Path function

Path functions are properties that are depend on the path followed during a process work(w) and Heat(q)

9.2.2 Thermodynamic law

• The first law

The first law of thermodynamics is the law of conservation of energy, which means that the energy can not be created or destroyed but converted from one form to another

• The second law

The second las of thermodynamics is the law of entropy, which means that total entropy of isolated system increases in spontaneous process.

• Gibbs free energy

Gibbs free energy helps us to determine the spontaneity of a reaction from the thermodynamic data of system

$$\Delta G = \Delta H - T \Delta S$$

Entropy change of universe	Gibbs free energy($\Delta G = -T\Delta S$(universe))	Direction of reaction
> 0	< 0	forward reaction
$= 0$	$= 0$	equilibrium
< 0	> 0	reverse reaction

Enthalpy ($\Delta Hsys.$) Entropy($\Delta Ssys.$)	positive (+) : endothermic	negative (−) : exothermic
positive(+)	At high temperature, forward reaction is spontaneous	Always forward reaction is spontaneous
negative(−)	Always forward reaction is nonspontaneous	At low temperature, forward reaction is spontaneous

• The third law

The entropy of a perfect crystalline substance is zero at 0K

Chapter Review Questions   9. Thermodynamics

[161-163] **Refer to the following**

(A) Heat of formation   (B) Heat of neutralization

(C) Heat of combustion   (D) Heat of vaporization

(E) Heat of dissolution

161 The enthalpy change of the reaction $C_3H_8(g) + 5O_2(g) \rightarrow 3CO_2(g) + 4H_2O(l)$ is

162 The enthalpy change of the reaction $H^+(aq) + OH^-(aq) \rightarrow H_2O(l)$ is

163 The enthalpy change of the reaction $2C(s, \text{graphite}) + 3H_2(g) \rightarrow C_2H_6(g)$ is

[164-165] **Refer to the following heating curve of solid**

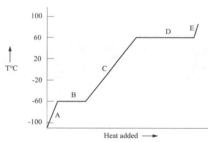

164 The region which shows the highest specific heat capacity

165 The region where all heat added is used to melting

166 The below is the standard enthalpy of formation(ΔH_f^o) of some materials.

Compound	Standard enthalpy of formation(ΔH_f^o, kJ/mol)
$C_2H_4(g)$	52
$H_2O(g)$	-241
$CO_2(g)$	-394

Which of the following is correct for the combustion reaction of $C_2H_4(g)$?

I.   When 2 mol of ethene is completely reacted with oxygen, 1322 kJ of heat is released.

II. The sum of bond dissociation energy of the reactants is greater than the product.

III. If the product is water rather than water vapor, more heat is released.

(A) I only   (B) II only   (C) III only   (D) II and III only   (E) I, II, and III

167 $CO_2(g)$ can be dissolved in water. Which of the following is correct for the process?

I.   The entropy change is positive

II. The enthalpy change is negative

III. The process is spontaneous at low temperature

(A) I only   (B) II only   (C) III only   (D) I, II only   (E) II, III only

168 Which of the following is correct for the reaction between calcium carbonate with hydrochloric acid?
(The heat of reaction, $\Delta H < 0$)

I.   The produced gas is acid anhydride

II. The temperature of surrounding increase when forward reaction proceeds

III. The forward reaction is always spontaneous

(A) I only   (B) II only   (C) I, II only   (D) II, III only   (E) I, II and III

169 200g of liquid is heated from 20℃ to 50℃ The liquid absorbed 2400cal. What is the specific heat of this liquid?

(A) 0.2cal/g℃   (B) 0.4cal/g℃   (C) 0.6cal/g℃   (D) 0.8cal/g℃   (E) 1cal/g℃

170 Which of the following is not correct?

(A) The temperature of surrounding decreases in endothermic reaction.

(B) Bond energy of reactants is larger than products in endothermic reaction.

(C) The sign of enthalpy change is negative in exothermic reaction.

(D) The system absorbed heat energy from surrounding in endothermic reaction.

(E) The temperature of surrounding increases when water is vaporized.

171 What is the final temperature of water when heated 20g of solid piece is immersed into 100mL of water. The conditions are below(solid piece is not soluble in water)

Initial temp. of solid piece	100℃
The specific heat of solid piece	0.5cal/g℃
Initial temperature of water	56℃
The specific heat of water	1cal/g℃
The density of water	1g/mL

(A) 30℃   (B) 40℃   (C) 50℃   (D) 60℃   (E) 70℃

172 The below is the result of calorimetry experiment

The volume of H_2O : 0.3L The density of water : 1g/mL The specific heat of water : 1cal/g℃ Initial temperature of H_2O : 20℃ Final temperature of H_2O : 60℃ Heat of combustion of C_3H_8 : 48,000cal/mol

What is the mass of C_3H_8 burned?

(A) 11g   (B) 22g   (C) 33g   (D) 44g   (E) 88g

173 Which of the following is not correct for the following reaction?

$2H_2(g) + O_2(g) \rightarrow 2H_2O(g)$ $\Delta H = -188kJ$

(A) The forward reaction is exothermic
(B) When the forward reaction proceeds, the temperature of surrounding increases
(C) When the forward reaction proceeds the entropy increase because the reaction is spontaneous at room temperature
(D) The product is more stable than reactants
(E) The sum of bond enthalpy of product is larger than that of reactants

174 The below is a thermochemical equation for several reactions.

$C(s, graphite) + 2H_2O(g) \rightarrow CO_2(g) + 2H_2(g)$ $\Delta H = 88kJ/mol$ $2H_2(g) + O_2(g) \rightarrow 2H_2O(g)$ $\Delta H = -482kJ/mol$

Which of the following is correct for the heat of reaction of the following reaction?

$C(s, graphite) + O_2(g) \rightarrow CO_2(g)$

(A) 324/mol   (B) −324kJ/mol   (C) 329kJ/mol   (D) −329J/mol   (E) −394kJ/mol

175 The below is thermochemical equations for several reactions.

$N_2(g) + O_2(g) \rightarrow 2NO(g)$ $\Delta H = +180kJ$ $2NO(g) + O_2(g) \rightarrow 2NO_2(g)$ $\Delta H = -113kJ$ $2N_2O(g) \rightarrow 2N_2(g) + O_2(g)$ $\Delta H = -163kJ$

Which of the following is not correct for the reaction?(The molar mass N : 14g/mol, O : 16g/mol)

(A) The standard enthalpy of formation of NO(g) is 90kJ/mol

(B) When gaseous covalent bond is broken, the bond energy is absorbed.

(C) When 1mol of N_2O(g) reacts with 1mol of NO_2(g) completely and only NO(g) is produced, 155kJ of heat is absorbed

(D) When 0.4mol of N_2(g) is completely converted to NO(g) 72kJ of heat is absorbed

(E) If 45g of NO(g) react with 48g of O_2(g), the theoretical yield of NO_2(g) is 92g and the amount of heat released is 169.5kJ

176 Which of the following is the process in increase of entropy?

(A) Freezing of water

(B) Deposition of I_2(g)

(C) N_2(g) react with O_2(g) to form a N_2O_5(g)

(D) Electrolysis of H_2O(l) at room temperature

(E) Ionic compound is formed from cation and anion

177 Which of the following is correct for the thermodynamic data?

(A) Endothermic reaction always spontaneous

(B) Endothermic reaction can be spontaneous at low temperature when entropy change value is negative

(C) Endothermic reaction always spontaneous at high temperature regardless of other thermodynamic data

(D) Exothermic reaction is always spontaneous when entropy change value is positive

(E) Exothermic reaction can be spontaneous at high temperature when entropy change value is negative

178 The below is the bond dissociation enthalpy of chemical bond

$N-H$ 393kJ/mol	$C=O$ 799kJ/mol	$H-O$ 467kJ/mol
$H-H$ 436kJ/mol	$N\equiv N$ 941kJ/mol	$N=N$ 418kJ/mol

Which of the following is correct for the enthalpy of formation of NH_3(g) from N_2(g) and H_2(g)?

(A) 54.5kJ/mol   (B) -54.5kJ/mol   (C) 109kJ/mol   (D) -109kJ/mol   (E) -218kJ/mol

179 Which of the following is different sign of entropy change in forward reaction?

(A) C_2H_6O(l) $+ 3O_2$(g) $\rightarrow 2CO_2$(g) $+ 3H_2O$(l)

(B) N_2(g) $+ 3H_2$(g) $\rightarrow 2NH_3$(g)

(C) HCl(g) $+ NH_3$(g) $\rightarrow NH_4Cl$(s)

(D) $2Fe_2O_3$(s) $+ 3C$(s) $\rightarrow 4Fe$(s) $+ 3CO_2$(g)

(E) $Ca(OH)_2$(aq) $+ CO_2$(g) $\rightarrow H_2O$(l) $+ CaCO_3$(s)

180 Which of the following is not correct for the represented reaction below?

$N_2O_4(g) \rightarrow 2NO_2(g)$

compound	ΔH_f^o(kJ/mol)	S^o(J/mol · K)
$N_2O_4(g)$	9	304
$NO_2(g)$	33	240

(A) At high temperature the forward reaction is spontaneous

(B) The forward reaction is nonspontaneous at 300K

(C) When the forward reaction proceeds the temperature of surrounding decreases

(D) The standard enthalpy of decomposition of $N_2O_4(g)$ is 57kJ

(E) When 1mol of $N_2O_4(g)$ is converted to $NO_2(g)$ completely, the standard entropy change is 176J/K

10   Chemical Kinetics

10.1   The Rate of Reaction

10.1.1 Expression of rate of reaction

• Average rate of reaction

The average rate of reaction is the rate in several time interval

$$aA(g) + bB(g) \rightarrow cC(g) + dD(g) \qquad rate = -\frac{1}{a}\frac{\Delta[A]}{\Delta t} = -\frac{1}{b}\frac{\Delta[B]}{\Delta t} = \frac{1}{c}\frac{\Delta[C]}{\Delta t} = \frac{1}{d}\frac{\Delta[D]}{\Delta t}$$

Generally average rate of reaction decreases as the reaction proceeds because as the reaction goes forward, there are fewer collisions between reactant molecules in other word, the concentration of reactant decreases

• Instantaneous rate of reaction

The instantaneous rate is the rate at any point. The instantaneous rate at time zero is initial rate of reaction

$$aA(g) + bB(g) \rightarrow cC(g) + dD(g) \qquad rate = -\frac{1}{a}\frac{d[A]}{dt} = -\frac{1}{b}\frac{d[B]}{dt} = \frac{1}{c}\frac{d[C]}{dt} = \frac{1}{d}\frac{d[D]}{dt}$$

10.1.2 Collision Theory

• Activation energy and activated complex at transition state

Activation energy is the minimum energy required to initiate a chemical reaction. After reaction initiates, the species are converted into activated complex temporarily at transition state before they form the product

• Collision theory and effective collision

The rate of reaction is affected by the collision number of reactants per unit time. Effective collision is the collision that can causes a chemical reaction. For the reaction, collisions with sufficient energy and proper orientation are needed

10.1.3 Factors affecting the rate of reaction

• Concentration

Concentration of reactants increases	→	Increase of Effective collisions	→	Number of particles(reactants) more than activation energy increases	→	Reaction rate increases

• Surface area

surface area of reactants increases	→	Increase of Effective collisions	→	Number of particles(reactants) more than activation energy increases	→	Reaction rate increases

• Pressure(only gaseous reaction)

Partial pressure of reactants increases	→	Increase of Effective collisions	→	Number of particles(reactants) more than activation energy increases	→	Reaction rate increases

• Temperature

Temperature of system increases	→	Increase of Effective collisions	→	Number of particles(reactants) more than activation energy increases	→	Reaction rate increases

Temperature of system increases	→	Number of particles(reactants) more than activation energy increases	→	Reaction rate increases

• Catalyst

Catalyst increase the rate of reaction through providing alternative reaction pathway but is not consumed in chemical reaction.

10.2   The Rate Law

10.2.1 The Rate Law for Elementary Reaction

- Rate law is expressed for the gaseous and aqueous solution
- Pure solvent and solid is not expressed in rate law but catalyst can be expressed in rate law
- The rate law can be expressed by molecularity in elementary reaction

Elementary reaction	Rate law	Unit of rate constant	Reaction order
A → Products	k[A]	1/s	first order
A + A → Products (2A → Products)	k[A]^2	1M/s	second order
A + B → Products	k[A][B]	1/Ms	second order
A + A + B → Products (2A + B → Products)	k[A]^2[B]	1/M^{2}s	third order
A + B + C → Products)	k[A][B][C]	1/M^{2}s	third order

k : rate constant

10.2.2 Experimental Determination of Rate Law

- Table from the experimental result

2NO(g) + 2H$_2$(g) → N$_2$(g) + 2H$_2$O(g)			
Experiment	[NO](M)	[H$_2$](M)	Initial rate(M/s)
1	5.0×10^{-3}	2.0×10^{-3}	1.3×10^{-5}
2	10.0×10^{-3}	2.0×10^{-3}	5.0×10^{-5}
3	10.0×10^{-3}	4.0×10^{-3}	10.0×10^{-5}

from the experiment rate = k[NO]^2[H$_2$]

- Reaction mechanism from the experimental result

Reaction mechanism is the series of elementary reaction. Rate can be expressed through rate determining step which is the slowest reaction of overall reaction. The activation energy is the largest in rate determining step

※ Intermediate : the molecules that formed in the preceding step and finally consumed in overall reaction

step 1 $H_2O_2(aq) + I^-(aq) \rightarrow H_2O(l) + IO^-(aq)$ slow
step 2 $H_2O_2(aq) + IO^-(aq) \rightarrow H_2O(g) + O_2(g) + I^-(aq)$ fast
overall reaction $2H_2O_2(aq) \rightarrow 2H_2O(l) + O_2(g)$

r ate = k[H$_2$O$_2$][I$^-$]   second order reaction   I$^-$: catalyst   IO$^-$: intermediate

10.2.3 Half-Life($t_{1/2}$) of first order reaction

The half-life($t_{1/2}$) is the time needed for concentration of reactant decrease to half of its initial concentration. The half-life of first order reaction is constant regardless of concentration of reactant. radio-decay reaction is the example of first order reaction

Chapter Review Questions — **10. Chemical Kinetics**

[181-182] **Refer to the following**

(A) Average rate of reaction
(B) Potential energy
(C) Activation energy
(D) Catalyst
(E) Effective collision

181 is the collision with sufficient energy and proper orientation that can causes a chemical reaction

182 is the minimum energy required to initiate the chemical reaction

183 Generally the reaction rate is the highest at the beginning of the reaction

Because

The concentration of reactants decreases as reaction is proceeding

184 Which of the following is correct?

I.   When the concentration of the reactants increases, the reaction rate can increases without change of average kinetic energy
II. When the temperature increases, reaction rate increases in the endothermic reaction but not in the exothermic reaction
III. When the reaction rate increases, the enthalpy change decreases

(A) I only   (B) II only   (C) III only   (D) I, II only   (E) II, III only

185 Which of the following is correct for the reaction rate?

I.   The rate of the gas evolved reaction can be determined by measuring the change in gas volume produced over a period of time.
II. As the temperature increases, the reaction rate increases because of the change in activation energy.
III. As the amount of reactant increases, the reaction rate increases.

(A) I only   (B) II only   (C) III only   (D) I, II only   (E) II, III only

186 Which of the following is correct for the determining the reaction rate?

    I. The rate raw can be obtained from experiment.

    II. Overall reaction rate can be determined by slow step in reaction mechanism.

    III. The activation energy of slow step is higher than other steps in mechanism.

    (A) I only   (B) II only   (C) III only   (D) I, II only   (E) I, II and III

187 Which of the following is correct for the first order reaction?

    I. The rate is proportional to the concentration of reactants

    II. The half$-$life is proportional to the concentration of reactants

    III. The rate constant is proportional to the concentration of reactants

    (A) I only   (B) II only   (C) III only   (D) I, II only   (E) I, II and III

188 Which of the following is correct for the activation energy of reverse reaction when the heat of reaction is $-50kJ$ and activation energy of forward reaction is 30kJ?

    (A) 10kJ   (B) 20kJ   (C) 70kJ   (D) 80kJ   (E) 100kJ

189 Which of the following is not correct for the elementary reaction represented below?

$$A + B \rightarrow C + D$$

    (A) There is no intermediate in the reaction

    (B) There is no catalysis in the reaction

    (C) The reaction order for [A] is first

    (D) The overall reaction order is second

    (E) When the concentration of [A] is doubled the rate increase 4 times faster than before

190 The below is the reaction between a piece of calcium carbonate and excess of hydrochloric acid

$$CaCO_3(s) + 2HCl(aq) \rightarrow CaCl_2(aq) + CO_2(g) + H_2O(l) \qquad \Delta H < 0$$

Which of the following is not correct for the reaction?

    (A) When calcium carbonate is reacted in powder form, the reaction rate increases.

    (B) The reaction rate increases with increasing temperature

    (C) When the concentration of HCl(aq) increases the volume of $CO_2(g)$ produced increases

    (D) When HCl(aq) disappears at a rate of 0.1M/s, the rate of formation of $CO_2(g)$ is 0.05M/s

    (E) The reaction rate is not affected by the mass of calcium carbonate

191 Which of the following is correct for the below table?

[A]	[B]	rate(M/s)
1	2	4×10^{-2}
2	2	8×10^{-2}
0.5	4	8×10^{-2}

	rate law	rate constant
(A)	$k[A][B]$	$2 \times 10^{-2} M \cdot s$
(B)	$k[A][B]^2$	$1 \times 10^{-2} M^{-2} \cdot s^{-1}$
(C)	$k[A][B]^3$	$1 \times 10^{-2} M^2 \cdot s^{-1}$
(D)	$k[A]^2[B]$	$2 \times 10^{-2} M \cdot s^{-2}$
(E)	$k[B]^2$	$2 \times 10^{-2} M \cdot s^{-1}$

192 Which of the following would be best represent the rate law, overall reaction, and intermediate of the chemical reaction?

Step 1 : $A + B \rightarrow C$	Slow
Step 2 : $C + B \rightarrow E$	Fast

	rate law	overall reaction	intermediate
(A)	$k[A][B]$	$A + B \rightarrow E$	A
(B)	$k[A][B]^2$	$A + B \rightarrow E$	B
(C)	$k[A][B]$	$A + 2B \rightarrow E$	C
(D)	$k[A][B]$	$A + 2B \rightarrow E$	B
(E)	$k[C][B]^2$	$A + 2B \rightarrow E$	E

193 The half-life of a reaction is independent of the concentration of the reactants. What is the order of reaction and unit of rate constant?(The unit of rate is M/s and the unit of concentration is M(molarity))

	order of reaction	unit of rate constant
(A)	zero	$M^{-1} \cdot s^{-1}$
(B)	first	$M \cdot s^{-1}$
(C)	first	s^{-1}
(D)	second	$M^{-1} \cdot s^{-1}$
(E)	second	s^{-1}

194 $A(g) \rightarrow B(g)$: first order reaction for the reactant A

The concentration of A before the reaction is 2.0M. When the reaction proceeded for 60 minutes, the concentration of A is 0.125 M, which of the following is correct for the half-life and the mole ratio between A and B(A : B)?

	half−life	mol ratio(A : B)
(A)	10min	1 : 7
(B)	15min	1 : 7
(C)	15min	1 : 15
(D)	20min	1 : 7
(E)	20min	1 : 15

195 Which of the following is related with catalyst?

(A) Enthalpy change and potential energy of the products

(B) Entropy change and potential energy of the product

(C) Reaction pathway and potential energy of the activated complex

(D) Enthalpy change and rate of forward reaction

(E) Enthalpy change and potential energy of the reactants

196 Which of the following is not correct for the chemical kinetics?

(A) Catalyst is not consumed in chemical reaction

(B) Reaction mechanism is the series of elementary reaction

(C) Activated complex is the substance temporarily existed at transition state

(D) The quantity of pure solid do not affect the rate of reaction

(E) The surface area of solid reactants do not affect the rate of reaction

197 Which of the following is not correct?

(A) The energy of reactants of collision affects the reaction rate

(B) The reaction rate can be decreased as reaction is proceeding

(C) When temperature increases forward reaction rate increase but reverse reaction rate remains constant

(D) The orientation of collision is important in chemical reaction

(E) Catalysis can lower the activation energy

198 Which of the following would most likely be associated with a reaction that occurs at a high rate?

(A) A large activation energy

(B) A small activation energy

(C) A small equilibrium constant

(D) A large equilibrium constant

(E) A negative Gibbs free energy

199 Which of the following is correct for the represented below?

$$CO_2(g) + H_2(g) \rightarrow CO(g) + H_2O(g) \qquad \Delta H = -41.2kJ$$

(A) When the temperature increases, the rate of forward reaction increases because activation energy decreases

(B) When the temperature increases, the rate of forward reaction decreases because the forward reaction is exothermic

(C) When the temperature decreases, the rate of forward reaction decreases because the number of reactants more than activation energy decreases

(D) When the catalyst is used, the rate of forward reaction increases because the heat of reaction decreases

(E) When the catalyst is used, the rate of forward reaction increases because the rate of reverse reaction decreases

200 The below is the results of rate measurement for the reaction $2NO(g) + O_2(g) \rightarrow 2NO_2(g)$

Experiment	Initial concentration of [NO]	Initial concentration of [O$_2$]	Initial rate of NO$_2$ formation(M/s)
1	0.20	0.20	4.0×10^{-2}
2	0.20	0.10	2.0×10^{-2}
3	0.40	0.10	8.0×10^{-2}

Which of the following is correct for the reaction?

(A) The forward reaction is endothermic

(B) The reaction order for NO(g) is first order

(C) The reaction order for O$_2$(g) is second order

(D) The unit of rate constant(k) is M^2s^{-1}

(E) When the concentrations of $[NO_2] = 0.10M$ and $[O_2] = 0.20M$, the initial rate of NO$_2$ formation is 1.0×10^{-2}

11 Chemical Equilibrium

11.1.1 Physical equilibrium

The physical equilibrium is the equilibrium of two different phases without changing of chemical composition. phase change is related with physical equilibrium. At physical equilibrium, the transition rates between two phases are the same

11.1.2 Chemical equilibrium

When the rate of forward reaction is the same as reverse reaction in chemical reaction, in other word, at equilibrium there is no change of concentration of reactants and products with time

• Dynamic equilibrium

The rate of transition rate of reactants and products are equal, which means there is microscopic change but no net change in two substances. Both physical and chemical equilibriums are dynamic equilibriums

11.2 The Expression of Equilibrium Constant

11.2.1 The law of mass action

$$aA\,(aq) + bB\,(aq) \rightleftharpoons cC\,(aq) + dD\,(aq) \qquad K_c = \frac{[C]^c[D]^d}{[A]^a[B]^b}$$

• Equilibrium constant(K) is only dependent to temperature
• Including the catalyst, pure solvent and solid are not contained in the equation

$$N_2(g) + 3H_2(g) \rightleftharpoons 2NH_3(g) \qquad K_C = \frac{[NH_3]^2}{[N_2][H_2]^3}$$

$$2N_2(g) + 6H_2(g) \rightleftharpoons 4NH_3(g) \qquad (K_C)' = (K_C)^2 = \frac{[NH_3]^4}{[N_2]^2[H_2]^6} = \left(\frac{[NH_3]^2}{[N_2][H_2]^3}\right)^2$$

$$\frac{1}{2}N_2(g) + \frac{3}{2}H_2(g) \rightleftharpoons NH_3(g) \qquad (K_C)'' = (K_C)^{\frac{1}{2}} = \frac{[NH_3]}{[N_2]^{\frac{1}{2}}[H_2]^{\frac{3}{2}}} = \left(\frac{[NH_3]^2}{[N_2][H_2]^3}\right)^{\frac{1}{2}}$$

$$2NH_3(g) \rightleftharpoons N_2(g) + 3H_2(g) \qquad K''' = (K_C)^{-1} = \left(\frac{[N_2][H_2]^3}{[NH_3]^2}\right) = \left(\frac{[NH_3]^2}{[N_2][H_2]^3}\right)^{-1}$$

$$H_2CO_3(aq) \rightleftharpoons H^+(aq) + HCO_3^-(aq) \qquad K_1 = \frac{[H^+][HCO_3^-]}{[H_2CO_3]}$$

$$HCO_3^-(aq) \rightleftharpoons H^+(aq) + CO_3^{2-}(aq) \qquad K_2 = \frac{[H^+][CO_3^{2-}]}{[HCO_3^-]}$$

$$H_2CO_3(aq) \rightleftharpoons 2H^+(aq) + CO_3^{2-}(aq) \qquad K = K_1 \times K_2 = \frac{[H^+][HCO_3^-]}{[H_2CO_3]} \times \frac{[H^+][CO_3^{2-}]}{[HCO_3^-]} = \frac{[H^+]^2[CO_3^{2-}]}{[H_2CO_3]}$$

11.2.2 Homogeneous and heterogeneous equilibrium

• Homogeneous equilibrium

All the species related with equilibrium are in the same phase. For example all the species are gas or aqueous solution

$$CH_3COOH(aq) + H_2O(l) \rightleftharpoons H_3O^+(aq) + CH_3COO^-(aq) \qquad CO(g) + Cl_2(g) \rightleftharpoons COCl_2(g)$$

• Heterogeneous equilibrium

There are two or more phases exist in chemical equilibrium

$$CaCO_3(s) \rightleftharpoons CaO(s) + CO_2(g) \qquad 2H_2O(g) + C(s) \rightleftharpoons CO_2(g) + 2H_2(g)$$

11.2.3 Types of equilibrium constant

• K_C

The equilibrium constant expressed with equilibrium constant, molarity

$$2SO_2(g) + O_2(g) \rightleftharpoons 2SO_3(g) \qquad K_C = \frac{[SO_3]^2}{[SO_2]^2[O_2]}$$

- K_P

  The equilibrium constant expressed with partial pressure

  $$2SO_2(g) + O_2(g) \rightleftharpoons 2SO_3(g) \qquad K_P = \frac{P_{SO_3}^2}{P_{SO_2}^2 P_{O_2}}$$

- K_{SP}

  K_{SP} is the solubility product (constant) is the product of the constituent ions

  $$AgCl(s) \rightleftharpoons Ag^+(aq) + Cl^-(aq) \qquad K_{SP} = [Ag^+][Cl^-]$$

- K_a and K_b

  K_a is acid ionization constant, and K_b is base ionization constant

  $$HC_2H_3O_2(aq) + H_2O(l) \rightleftharpoons H_3O^+(aq) + C_2H_3O_2^-(aq) \qquad K_a = \frac{[H_3O^+][C_2H_3O_2^-]}{[HC_2H_3O_2]}$$

  $$NH_3(aq) + H_2O(l) \rightleftharpoons NH_4^+(aq) + OH^-(aq) \qquad K_b = \frac{[NH_4^+][OH^-]}{[NH_3]}$$

11.2.3 Types of equilibrium constant

- K value and favored reaction

$K > 1$ Products are favored than reactants at equilibrium
$K = 1$ Products and reactants are equally favored at equilibrium
$K < 1$ Reactants are favored than products at equilibrium

- Reaction quotient(Q_C) and prediction the direction of the reaction

  The reaction quotient(Q_C) is expressed as law of mass action but the equilibrium concentrations are substituted by initial concentrations at certain point or experimental condition

$Q < K$ To reach the equilibrium, reactants are converted into products : The reaction shifts from reactants to products
$Q = K$ The system is at equilibrium. there is no net shift of reaction
$Q > K$ To reach the equilibrium, products are converted into reactants : The reaction shifts from products to reactants

11.3   Factors that Affect the Chemical Equilibrium : Le Chatelier's Principle

11.3.1 Le Chatelier's Principle

When stress is applied to the equilibrium, the system reduces the stress added to the system through a chemical reaction to reach a new equilibrium position. The change of concentration, partial pressure, volume and temperature can be the stress

11.3.2 Factors that affect the chemical equilibrium

• Concentration   No change in equilibrium constant

When the reactant is added or product is removed at the equilibrium state, the system will shift to new equilibrium state through forward reaction. When the product is added or reactant is removed at the equilibrium state, the system will shift to new equilibrium state through reverse reaction.

Change in concentration at equilibrium state	Direction of shift	Equilibrium constant
Concentration of reactant increases	Right side, forward reaction	No change
Concentration of reactant decreases	Left side, reverse reaction	No change
Concentration of product increases	Left side, reverse reaction	No change
Concentration of product decreases	Right side, forward reaction	No change

• Volume   No change in equilibrium constant

When the total pressure is increased by decreasing the volume at equilibrium state, the system will shifts to the direction of decreasing the total number of gases. When the total pressure is decreased by increasing volume at equilibrium state, the system shifts to the direction of increasing the total number of gases.

Change in volume at equilibrium state	Direction of shift	Equilibrium constant
Increasing of volume(total pressure decreases)	Direction of increasing the total number of gases	No change
Decreasing of volume(total pressure increases)	Direction of decreasing the total number of gases	No change

• Adding noble gas   No change in equilibrium constant

When noble gas is added to the rigid container at equilibrium state, there is no shift of equilibrium because there is no effect on the equilibrium partial pressure ratio. When noble gas is added to equilibrium under constant pressure conditions, the system shifts to the direction of increasing the total number of gases because it is the same effect of decreasing the total pressure by increasing the volume

Adding noble gas at equilibrium state	Direction of shift	Equilibrium constant
At constant volume	No shift	No change
At constant pressure	Direction of increasing the total number of gases	No change

• Temperature   Change in equilibrium constant

For endothermic reaction, when temperature increases at equilibrium states, the equilibrium will shift the right and equilibrium constant(K) increases. and when temperature decreases at equilibrium states, the equilibrium will shits the left and equilibrium constant(K) decreases. For exothermic reaction, when temperature increases at equilibrium states, the equilibrium will shift the left and equilibrium constant(K) decreases. and when temperature decreases at equilibrium states, the equilibrium will shits the right and equilibrium constant(K) increases.

Change in temperature at equilibrium state	Direction of shift	Equilibrium constant
Increasing the temperature of endothermic reaction	Right side, forward reaction	Increases
Decreasing the temperature of endothermic reaction	Left side, reverse reaction	Decreases
Increasing the temperature of exothermic reaction	Left side, reverse reaction	Decreases
Decreasing the temperature of exothermic reaction	Right side, forward reaction	Increases

Chapter Review Questions   **11. Chemical Equilibrium**

[201-202] **Refer to the following**

(A) Dynamic equilibrium   (B) Rate constant

(C) Activation energy   (D) Solubility of solute

(E) Equilibrium constant

201 is increases when the temperature increases

202 states the situation that rate of forward reaction is the same as reverse reaction

203 At equilibrium, the concentration of reactants and products remain constant

Because

At equilibrium, the molarity of reactants and products are the same

204 The reaction with a large activation energy has a small equilibrium constant

Because

The larger the activation energy, the slower the reaction rate

205 When sodium acetate, $NaC_2H_3O_2$, is added to the equilibrium mixture represented by the equation below, the concentration of H_3O^+ decreases

$$HC_2H_3O_2 + H_2O \rightleftharpoons H_3O^+ + C_2H_3O_2^-$$

Because

The equilibrium constant of a reaction changes as the concentration of the reactants changes

206 Which of the following is(are) dependent with temperature?

I. K_c(equilibrium constant as molarity)

II. K_{sp}(equilibrium constant of insoluble sate)

III. k(rate constant)

IV. Molarity

(A) I only   (B) I and II only   (C) I and III only

(D) III and IV only   (E) I, II, III, and IV

207 Which of the following is correct?

   I. At equilibrium, the concentration of reactants and products remain constant

   II. At equilibrium, the concentration of reactants are the same as that of products

   III. At equilibrium, forward and reverse reaction does not occur

(A) I only   (B) II only   (C) III only   (D) I, II only   (E) II and III only

208 Which of the following is correct for the below reaction?

$$Fe(OH)_3(s) \rightleftharpoons Fe^{3+}(aq) + 3OH^-(aq) \quad K_{sp} = 2 \times 10^{-39}$$

   I. $K_{sp} = [Fe^{3+}][OH^-]^3$

   II. Solubility increase at low pH

   III. $[Fe^{3+}] = [OH^-]$ at equilibrium

(A) I only   (B) II only   (C) III only   (D) I, II only   (E) II and III only

209 Which of the following is correct?

$$2NO_2(g) \rightleftharpoons N_2O_4(g) \quad K = 2.5$$

   I. The reaction lies to forward reaction

   II. The equilibrium constant of reverse reaction is 0.4

   III. The equilibrium constant of $3N_2O_4(aq) \rightleftharpoons 6NO_2(aq)$ is 1.2

(A) I only   (B) II only   (C) III only   (D) I and II only   (E) I, II, and III

210 Which of the following is correct for the reaction below?

$$CH_4(g) + H_2O(g) + energy \rightleftharpoons CO(g) + 3H_2(g)$$

   I. Increasing the temperature can increase the yield of the product.

   II. Periodic recovery of the product can increase the yield of the product.

   III. Since $H_2O(g)$ is a solvent, it is not included in the equilibrium equation.

(A) I only   (B) II only   (C) III only   (D) I and II only   (E) I, II, and III

211 Which of the following is correct for the reaction below?

$$N_2(g) + 3H_2(g) \rightleftharpoons 2NH_3(g) + energy$$

I. The equilibrium constant is higher at low temperature than at high temperature, but the reaction can not be occurred at low temperature

II. As the pressure increases, the equilibrium constant increases

III. The yield of the product can be increased at low temperature and high pressure.

(A) I only   (B) II only   (C) III only   (D) I and II only   (E) I and III only

212 Which of the following is correct expression of equilibrium constant for the reaction represented below?([ ] is the expression of molarity, P is the partial pressure of substance)

$$CaCO_3(s) \rightleftharpoons CaO(s) + CO_2(g)$$

(A) $K_C = \dfrac{[CaO] + [CO_2]}{[CaCO_3]}$ (B) $K_C = \dfrac{[CaCO_3]}{[CaO][CO_2]}$ (C) $K_C = \dfrac{[CaCO_3]}{[CO_2]}$

(D) $K_P = P_{CO2}$ (E) $K_P = \dfrac{1}{P_{CO2}}$

213

$$N_2(g) + \frac{1}{2}O_2(g) \rightleftharpoons N_2O(g) \qquad K_1$$

$$8NO_2(g) \rightleftharpoons 4N_2O_4(g) \qquad K_2$$

$$2N_2(g) + 4O_2(g) \rightleftharpoons 4NO_2(g) \qquad K_3$$

Which of the following is correct for the equilibrium constant for the reaction below?

$$2N_2O(g) + 3O_2(g) \rightleftharpoons 2N_2O_4(g) \quad K =$$

(A) $\dfrac{(K_2)^{\frac{1}{2}}(K_3)}{K_1^2}$ (B) $\dfrac{(K_1)(K_3)^{\frac{1}{2}}}{K_2^2}$ (C) $\dfrac{(K_2)(K_3)^{\frac{1}{2}}}{K_1^2}$

(D) $\dfrac{(K_1)\frac{1}{2}(K_3)}{K_2^2}$ (E) $\dfrac{(K_2)^2(K_3)}{K_1^2}$

214 Which of the following is correct for the equilibrium constant?

(A) The equilibrium constant for products favored reaction is lager than 1

(B) The equilibrium constant is not affected by the temperature change

(C) The reaction with large equilibrium constant reaches to equilibrium state very fast

(D) If the reaction with small equilibrium constant, there is no product at equilibrium state

(E) For the gaseous reaction, equilibrium constant can be changed by the total pressure

215 Which of the following is correct for the relationship between quotient(Q) and equilibrium constant(K) at constant temperature?

(A) When Q is larger than K, the reaction shift to reverse reaction and equilibrium constant decreases

(B) When Q is larger than K, the reaction shift to forward reaction and equilibrium constant does not change

(C) When Q is larger than K. the reaction shift to forward reaction and equilibrium constant increases

(D) When Q is smaller than K, the reaction shift to forward reaction and equilibrium constant does not change

(E) When Q is smaller than K, the reaction shift to forward reaction and equilibrium constant increases

216 Which of the following substances has the lowest solubility?

(A) $CaSO_4(s)$, $K_{sp} = 9.1 \times 10^{-6}$ (B) $PbCrO_4(s)$, $K_{sp} = 2.8 \times 10^{-13}$

(C) $AgCl(s)$, $K_{sp} = 1.8 \times 10^{-10}$ (D) $BaCO_3(s)$, $K_{sp} = 5.1 \times 10^{-9}$

(D) $ZnC_2O_4(s)$, $K_{sp} = 2.7 \times 10^{-10}$

217 The below reaction is at equilibrium

$$N_2(g) + 3H_2(g) \rightleftharpoons 2NH_3(g) \qquad \Delta H = -92.2 kJ$$

Which of the following would not shift to forward reaction?

(A) increasing the N_2 concentration
(B) increasing the pressure
(C) doubling the concentration of each substance, reactant and product
(D) increasing the temperature
(E) removing the $NH_3(g)$

218 Consider the following equilibrium system

$$N_2(g) + 3H_2(g) \rightleftharpoons 2NH_3(g)$$

If the total pressure of the system is increased, which of the following is possible?

(A) The concentration of $N_2(g)$ is not affected by total pressure
(B) The concentration of $H_2(g)$ is not affected by total pressure
(C) The concentration of $NH_3(g)$ is not affected by total pressure
(D) The reaction will shift to right
(E) The reaction will shift to left, so equilibrium constant is decreased

219 Initially the reaction represented below is at equilibrium state. Which of the following is correct condition to increase NO yield?

$$N_2(g) + O_2(g) + energy \rightleftharpoons 2NO(g)$$

(A) Decrease the $N_2(g)$ concentration
(B) Decrease the O_2 concentration
(C) Increase the temperature
(D) Increase the pressure
(E) Decrease the pressure

220 Which of the following is not correct for the reaction below?

$$CaCO_3(s) + energy \rightleftharpoons CaO(s) + CO_2(g)$$

(A) Entropy increases as the forward reaction proceeds
(B) When the temperature is increased, the forward reaction rate increases but the reverse reaction rate decreases.
(C) The amount of calcium carbonate is not affected by the position of the equilibrium.
(D) The equilibrium constant increases when the temperature is increased.
(E) The catalyst can not change the equilibrium constant but allows it to reach equilibrium quickly

12 Acid and Base

12.1 Acid and Base

12.1.1 Definition of Acid and Base

• **Arrhenius definition**

Arrhenius acid is substance that ionized in water to produce hydrogen ion, $H^+(aq)$. Arrhenius base is substance that ionized in water to produce hydroxide ion, $OH^-(aq)$

Arrhenius acid	Arrhenius base
$HCl(aq) \rightarrow H^+(aq) + Cl^-(aq)$ $HNO_3(aq) \rightarrow H^+(aq) + NO_3^-(aq)$ $H_2SO_4(aq) \rightarrow 2H^+(aq) + SO_4^{2-}(aq)$ $HC_2H_3O_2(aq) \rightleftharpoons H^+(aq) + C_2H_3O_2^-(aq)$	$NaOH(aq) \rightarrow Na^+(aq) + OH^-(aq)$ $Ca(OH)_2(aq) \rightarrow Ca^{2+}(aq) + 2OH^-(aq)$ $Mg(OH)_2(aq) \rightleftharpoons Mg^{2+}(aq) + 2OH^-(aq)$

• **Brønsted-Lowry definition**

Brønsted-Lowry acid is proton(H+)donor and Brønsted-Lowry base is proton(H+) acceptor. Conjugate acid-base concept is important in Brønsted-Lowry definition

$HCl(aq) + H_2O(l) \rightarrow H_3O^+(aq) + Cl^-(aq)$ $K_a \gg 1$ Acid 1 Base 2 Acid 2 Base 1	• Acid 1 − Base 1 are conjugate acid-base pair • Acid 2 − Base 2 are conjugate acid-base pair • $HCl(aq)$ is stronger acid than $H_3O^+(aq)$ • $H_2O(l)$ is stronger base than $Cl^-(aq)$
$HC_2H_3O_2(aq) + H_2O(l) \rightarrow H_3O^+(aq) + C_2H_3O_2^-(aq)$ $K_a \ll 1$ Acid 1 Base 2 Acid 2 Base 1	• Acid 1 − Base 1 are conjugate acid-base pair • Acid 2 − Base 2 are conjugate acid-base pair • $H_3O^+(aq)$ is stronger acid than $HC_2H_3O_2(aq)$ • $C_2H_3O_2^-(aq)$ is stronger base than $H_2O(l)$
$NH_3(aq) + H_2O(l) \rightarrow NH_4^+(aq) + OH^-(aq)$ $K_b \ll 1$ Base 1 Acid 2 Acid 1 Base 2	• Acid 1 - Base 1 are conjugate acid-base pair • Acid 2 - Base 2 are conjugate acid-base pair • $NH_4^+(aq)$ is stronger acid than $H_2O(l)$ • $OH^-(aq)$ is stronger base than $NH_3(aq)$

- Conjugate acid base pair differ only in the presence of a proton, $H^+(aq)$
- Amphoteric species can act as both acidic and basic, and they can react with acid or base for example, Al_2O_3 can react with both acid and base
- Amphiprotic can either accept or donate proton, $H^+(aq)$

$$OH^- \xleftarrow{-H^+} H_2O \xrightarrow{+H^+} H_3O^+$$

• **Lewis definition**

Lewis acid is electron pair acceptor and Lewis base is electron pair donor

$$\begin{array}{cccc}
& F & H & \\
& | & | & \\
F\!-\!B & & N\!-\!H & \longrightarrow \\
& | & | & \\
& F & H &
\end{array} \qquad
\begin{array}{cccc}
F & H \\
| & | \\
F\!-\!B\!-\!N\!-\!H \\
| & | \\
F & H
\end{array}$$

acid   base

12.1.2 Autoionization constant

• **Autoionization of water**

$H_2O(l) + H_2O(l) \rightleftharpoons H_3O^+(aq) + OH^-(aq)$ or $H_2O(l) \rightleftharpoons H^+(aq) + OH^-(aq)$

Ion product of water $Kw = [H_3O^+][OH^-] = 1.0 \times 10^{-14}(25℃)$

• **pKw, pH and pOH**

$pH = -\log[H_3O^+] = -\log[H^+]$ $pOH = -\log[OH^-]$ $Kw = [H^+][OH^-]$

$-\log Kw = pKw = -\log[H^+] + (-\log[OH^-]) = pH + pOH$ $(pKw = pH + pOH)$

at 25℃ $pH + pOH = 14$

12.1.3 pH Calculation of Acid and Base

• **pH of strong acid and strong base**

Strong acids are strong electrolytes which are ionized completely in water. For example, the pH of 0.1M $HCl(aq)$ is 1, because $HCl(aq)$ is ionized completely and the concentration of hydrogen ion, $[H^+]$ is 0.1M which is the same as the initial concentration of $HCl(aq)$. pH of strong base can be calculated from the pKw, and pOH. For example, the pOH of 0.1M NaOH is 1 and the pH is the 13 at 25℃ because $pH + POH = 14$ at 25℃

− strong acid : $HCl(aq)$, $HBr(aq)$, $HI(aq)$, $HNO_3(aq)$, $HClO_4(aq)$, $H_2SO_4(aq)$

− strong base : $NaOH(aq)$, $KOH(aq)$, $Ca(OH)_2(aq)$

• **pH of weak acid and weak base**

Most of acids and bases are weak acids and weak bases. They are only partially ionized in water. At equilibrium the solution of them contain both ionized and molecular form. Most are molecular form not ionized.

— weak acid : $HC_2H_3O_2(aq)$, $HCN(aq)$, $HF(aq)$

— weak base : $NH_3(aq)$, $CH_3NH_2(aq)$

$$pH = -\log \sqrt{K_aC_a} \qquad pOH = -\log \sqrt{K_bC_b}$$

pH of 0.1M $HC_2H_3O_2(aq)$ ($K_a = 10^{-5}$, 25℃) $pH = -\log \sqrt{10^{-5} \times 0.1} = 3$

pH of 0.1M $NH_3(aq)$ ($K_b = 10^{-5}$, 25℃) $\quad pOH = -\log \sqrt{10^{-5} \times 0.1} = 3 \quad pH = 14 - pOH = 14 - 3 = 11$

• **Strength of Acid and Acidity, Strength of Base and Basicity**

Strength of acid is determined by acid ionization constant, K_a, or degree of ionization(α) but acidity is determined by pH. The strength of acidity of $HCl(aq)$ is larger than $HC_2H_3O_2(aq)$ regardless of concentration. The Acidity of 0.1M of $HC_2H_3O_2$($K_a = 10^{-5}$) is larger than that of 10^{-4}M of $HCl(aq)$. The strength of base of $NaOH(aq)$ is larger than $NH_3(aq)$ regardless of concentration. The basicity of 0.1M of $NH_3(aq)$($K_b = 10^{-5}$) is larger than 10^{-4}M of $NaOH(aq)$.

12.2 Acid-Base Reaction : Neutralization

12.2.1 Equation of Neutralization

• **Strong acid and strong base**

$HCl(aq) + NaOH(aq) \rightarrow H_2O(l) + NaCl(aq)$ $\qquad$ pH = 7.0(25℃)
- Net ionic equation of neutralization : $H^+(aq) + OH^-(aq) \rightarrow H_2O(l)$
- Net ionic equation of overall reaction : $H^+(aq) + OH^-(aq) \rightarrow H_2O(l)$
- Spectator ion : $Na^+(aq)$, $Cl^-(aq)$

• **Strong acid and strong base**

$H_2SO_4(aq) + Ca(OH)_2(aq) \rightarrow 2H_2O(l) + CaSO_4(s)$ $\qquad$ pH = 7.0(25℃)
- Net ionic equation of neutralization : $H^+(aq) + OH^-(aq) \rightarrow H_2O(l)$
- Net ionic equation of overall reaction : $2H^+(aq) + SO_4^{2-}(aq) + Ca^{2+}(aq) + 2OH^-(aq) \rightarrow 2H_2O(l) + CaSO_4(s)$

• **Strong acid and weak base**

$HNO_3(aq) + NH_3(aq) \rightarrow NH_4NO_3(aq)$ $\qquad$ pH < 7.0(25℃)
- Net ionic equation of neutralization : $H^+(aq) + NH_3(aq) \rightarrow NH_4^+(aq)$
- Net ionic equation of overall reaction : $H^+(aq) + NH_3(aq) \rightarrow NH_4^+(aq)$
- Spectator ion : $NO_3^-(aq)$

- **Weak acid and strong base**

$HC_2H_3O_2(aq) + NaOH(aq) \rightarrow H_2O(l) + NaC_2H_3O_2(aq)$ $pH > 7.0(25℃)$

- Net ionic equation of neutralization : $HC_2H_3O_2(aq) + OH^-(aq) \rightarrow H_2O(l) + C_2H_3O_2^-(aq)$
- Net ionic equation of overall reaction : $HC_2H_3O_2(aq) + OH^-(aq) \rightarrow H_2O(l) + C_2H_3O_2^-(aq)$
- Spectator ion : $Na^+(aq)$

- **Weak acid and weak base**

$HC_2H_3O_2(aq) + NH_3(aq) \rightarrow NH_4C_2H_3O_2(aq)$ $HC_2H_3O_2, K_a = 10^{-5}, NH_3, K_b = 10^{-5}, pH = 7.0(25℃)$

- Net ionic equation of neutralization : $HC_2H_3O_2(aq) + NH_3(aq) \rightarrow NH_4C_2H_3O_2(aq)$
- Net ionic equation of overall reaction : $HC_2H_3O_2(aq) + NH_3(aq) \rightarrow NH_4C_2H_3O_2(aq)$

12.2.2 Titration and Indicator

- **Titration**

Acid-base titration is a volumetric analysis method that determines the unknown concentration of solution through the reaction between standard solution and solution with unknown concentration.

$n_aM_aV_a = n_bM_bV_b$

n_a : equivalence of acid M_a : molarity of acid V_a : volume of acid

n_a : equivalence of base M_b : molarity of base V_b : volume of base

- **Titrant**

Titrant is standard solution, a known concentration solution

- **Titrand**

Titrand is a solution, a unknown concentration solution. In titration, titrant reacts with a solution of analyte and the concentration of analyte can be determined.

- **Indicator**

Acid-base indicator is used to determine the endpoint in an acid-base titration. The endpoint of a titration occurs when the indicator changes color

Equivalent point	Endpoint
Equivalent point means the point at which equivalent amount of acids and bases are mixed	Endpoint means the point at which the color of indicator changes

Indicator	pH range	Acidic	Neutral	Basic
Methyl orange	4 − 6	orange	orange	yellow
phenolphthalein	9 − 10	colorless	colorless	pink
Bromthymol blue	6 − 7	yellow	green	blue

12.2.3 Buffer

• **Buffer**

Buffer solution refers to a solution in which the pH is maintained almost constant even when a small amount of strong acid or strong base is added. The buffer solution consists of a weak acid and its conjugate base or vice versa. When a small amount of strong base is added to the buffer solution, the weak acid reacts, and when a small amount of strong acid is added, the weak base reacts to adjust the pH change.

• **Preparation of buffer**

-Weak acid and its salt : $HC_2H_3O_2(aq) + NaC_2H_3O_2(aq)$

-Weak base and its salt : $NH_3(aq) + NH_4Cl(aq)$

-Weak acid is titrated with strong base incompletely : $HC_2H_3O_2(aq) + NaOH(aq) \rightarrow H_2O(l) + NaC_2H_3O_2(aq)$

-Weak base is titrated with strong acid incompletely : $NH_3(aq) + HCl(aq) \rightarrow NH_4Cl(aq)$

Chapter Review Questions 12. Acid and Base

[221-222] **refer to the following aqueous solutions**

(A) Weak acid
(B) Strong acid
(C) Brønsted−Lowry acid
(D) Brønsted−Lowry base
(E) Lewis acid

221 Always dissociates nearly completely in aqueous solution

222 K_a value is needed to calculate the pH of this solution

223 HF is strong base in aqueous solution

Because

HF can form a hydrogen bond with H_2O molecule

224 Hydrogen concentration of the aqueous solution with pOH 12 is higher than that of pOH 10 at 25℃

Because

pH + pOH = 14 at 25°C

225 Which of the following is(are) correct 0.1M of HCN(aq)?($K_a = 4.9 \times 10^{-10}$)

I. pH of HCN(aq) is 1 because HCN is strong acid
II. $[HCN] = [CN^-]$ at equilibrium
III. $[H^+] = [CN^-]$ at equilibrium
IV. $[HCN] > [CN^-]$ at equilibrium

(A) I and II only   (B) II and III only   (C) II and IV only
(D) III and IV only   (E) I, II, III and IV

226 Which of the following is correct for the neutralization between NH_3(aq) and HCN(aq) at 25℃

K_b of NH_3(aq) $= 1.8 \times 10^{-5}$	K_a of HCN(aq) $= 4.9 \times 10^{-10}$

I. NH_3(aq) is weak base
II. Salt from the neutralization is NH_4CN(aq)
III. pH at equivalent point is 7

(A) I only   (B) II only   (C) III only

(D) I and II only   (E) II, and III only

227 Which of the following is correct for making the buffer?

I. Weak acid and its conjugate base are mixed in water
II. Weak acid is titrated with strong base incompletely
III. Salt from weak acid with strong base is dissolved in water

(A) I only   (B) II only   (C) III only   (D) I, II only   (E) II, III only

228 Which of the following can act as both acid and base?

I. HCO_3^-
II. HPO_4^{2-}
III. CO_3^{2-}

(A) I only   (B) II only   (C) I, and II only

(D) II, and III only   (E) I, II, and III

229 Which of the following is correct for acid and base titration?

I. In titration of NaOH(aq) and HF(aq), the concentration of Na^+ is greater than F^- at the equivalence point
II. At $25\,^{\circ}C$, the sum of pK_a of HCN and pK_b of NH_3 is 14.
III. The degree of ionization of $HC_2H_3O_2$ is larger in $NaC_2H_3O_2$(aq) solution than in pure(distilled) water at $25\,^{\circ}C$
IV. Ionization constant of $HC_2H_3O_2$ is smaller in $NaC_2H_3O_2$(aq) solution than in pure(distilled) water at $25\,^{\circ}C$

(A) I only   (B) II only   (C) I, and II only

(D) II, and III only   (E) I, II, and III

230 K_a of weak acid HA $= 1.0\times10^{-5}$ at $25\,^{\circ}C$. Which of the following is correct for 0.1M HA (aq) solution

I. The concentration of A^- (aq) is the same as HA(aq)
II. pOH $= 11$
III. pK_b of conjugate base of HA is 1.0×10^{-9}

(A) I only   (B) II only   (C) I, and II only

(D) II, and III only   (E) I, II, and III

231 Which of the following is correct when 1.0 L of 0.1M ethanoic acid($HC_2H_3O_2$) is titrated with 0.05M sodium hydroxide(NaOH)? (K_a value of ethanoic acid : 1.0×10^{-5} at 25℃)

I. Before the titration the pH of $HC_2H_3O_2$(aq) is 1

II. In the titration process, a buffer solution forms before reaching the end point.

III. Phenolphthalein is a suitable indicator for identifying endpoints.

(A) I only   (B) II only   (C) I, and II only

(D) II, and III only   (E) I, II, and III

232 Which of the following is Lewis base?

(A) HNO_3 (B) NH_4^+ (C) $CH_3CH_2NH_2$ (D) $BeCl_2$ (E) BF_3

233 When 100mL of 0.10M HNO_3(aq) is diluted with distilled water to make 500mL of solution, which of the following is correct for the molarity of hydrogen ion?

(A) 0.0010M   (B) 0.0050M   (C) 0.010M   (D) 0.020M   (E) 0.05M

234 What is the volume of 2M of HCl for preparation of 0.5M of HCl solution 800mL ?

(A) 100mL   (B) 200mL   (C) 300mL   (D) 400mL   (E) 500mL

235 The below is the ionization reaction and ionization constant of acetic acid. Which of the following is not correct for the ionization of 0.05L of 10^{-2}M CH_3COOH

$$CH_3COOH(aq) + H_2O(l) \rightleftharpoons H_3O^+(aq) + CH_3COO^-(aq) \qquad K_a = 1.8 \times 10^{-5}$$

(A) pH of 10^{-2}M of acetic acid is higher than 2

(B) At equilibrium state the concentration of CH_3COOH(aq) is higher than that of H^+(aq)

(C) When 10^{-1}M of $CH_3COONa(NaC_2H_3O_2)$ is added the solution can be buffer

(D) The volume of 10^{-2}M of NaOH(aq) to neutralize acetic acid is smaller than 0.05L

(E) H_2O(l) is the base and weaker than CH_3COO^-(aq)

236 Which of the following aqueous solution would have a pH greater than 7 at 25℃?

(A) 10^{-8}M of CH_3COOH (B) 10^{-8}M of KNO_3 (C) 10^{-8}M of K_2CO_3

(D) 10^{-8}M of NH_4Cl (E) 10^{-8}M of HCl

237 How many milliliter of 0.2M calcium hydroxide must be added to a 400mL solution of 0.1M of acetic acid to complete the neutralization reaction and what is the pH at equivalent point at 25℃?

(A) 100mL, pH is 7   (B) 100mL, pH is smaller than 7

(C) 100mL, pH is larger than 7   (D) 200mL, pH is 7

(E) 200mL, pH is smaller than 7

238 Which of the following is not correct?

(A) Neutralization is exothermic reaction

(B) The pH of 10^{-2}M CH_3COOH(aq) is 2

(C) The net ion equation of neutralization between strong acid and strong base is H^+(aq) + OH^-(aq) → H_2O(l)

(D) The salt from neutralization between strong base and weak acid is basic

(E) pH changes very little when a small amount of strong acid or base is added to buffer

239 A titration experiment is conducted in which 15mL of a 0.015M $Ca(OH)_2$ solution is added to 30mL of an HCl solution of unknown concentration. Which of the following is the concentration of HCl(aq)?

(A) 0.015M   (B) 0.030M   (C) 0.045M   (D) 0.060M   (E) 0.075M

240 Which of the following is not correct for the titration of acid and base?(K_a(HCN) = 4.9×10^{-10})

(A) The volumes of 0.1M NaOH (aq) required to neutralize the same concentration and volume of HCl and $HC_2H_3O_2$ are the same.

(B) A solution of 100mL of 0.1M $HC_2H_3O_2$ and 100 mL of 0.05M NaOH (aq) is a buffer solution

(C) The aqueous solution of HCN and NaCN in a molar ratio of 1: 1 is a buffer solution.

(D) In acid and base titration, a small amount of indicator should be added to the standard solution(titrant).

(E) The pH of the end point is not always 7 in the acid base titration at a constant temperature of 25°C.

13 Electrochemistry

13.1 Oxidation and Reduction

13.1.1 Definition oxidation and reduction

	Oxidation	Reduction
Oxygen(O)	gain	lose
Hydrogen(H)	lose	gain
Electron(e^-)	lose	gain
Oxidation number	increase	decrease

• **Oxidizing agent**

The ion or molecule that accepts electrons(reduction) is called the oxidizing agent

• **Reducing agent**

The ion or molecule that donates electrons(oxidation) is called the reducing agent

Oxidation number	
1. oxidation number of an element is 0	
2. Alkali metal	$+1$
3. Alkaline earth metal	$+2$
4. Al(aluminium)	$+3$
5. F(fluorine)	-1
6. H(hydrogen)	$+1$
7. O(oxygen)	-2
8. Cl(chorine), Br(bromine), I(iodine)	-1

13.1.2 Activity series of metals

K>Ca>Na>Mg>Al>Zn>Fe>Ni>Sn>Pb>(H)>Cu>Hg>Ag>Pt>Au

- **Metal + Acid**

  $Zn(s) + 2HCl(aq) \rightarrow ZnCl_2(aq) + H_2(g)$

  $Cu(s) + 2HCl(aq) \rightarrow$ No reaction

- **Metal + Water**

  $2Na(s) + 2H_2O(l) \rightarrow 2NaOH(aq) + H_2(g)$

  $Ca(s) + 2H_2O(l) \rightarrow Ca(OH)_2(aq) + H_2(g)$

- **Metal + Metallic salt**

  $Zn(s) + Cu(NO_3)_2(aq) \rightarrow Zn(NO_3)_2(aq) + Cu(s)$

  $Cu(s) + Zn(NO_3)_2(aq) \rightarrow$ No reaction

13.1.3 Oxidation and reduction of halogen

Element	Color	chemical reactivity	color in CCl (l)
F_2	colorless		colorless
Cl_2	pale green	⇑ increase	pale green
Br_2	red-brown		brown
I_2	black-red		purple

$F_2(aq) + 2NaCl(aq) \rightarrow 2NaF(aq) + Cl_2(aq)$ $Cl_2(aq) + 2NaF(aq) \rightarrow$ No reaction

13.1.4 Balancing the Redox reaction

- **Procedure**

  - Write the unbalanced equation for the reaction in ionic form

  - Separate the equation into two half reaction(oxidation and reduction)

  - Balancing the mass (O : $H_2O(l)$, H : $H^+(aq)$)

  - Balancing the charge(electron : e^-)

  - If the oxidation reduction half reactions contains different numbers of electrons, multiplication is used to equalize the number of electrons

  - Overall reaction can be obtained through adding two half equation

- **Acidic and basic condition**

$Sn^{2+}(aq) + MnO_4^-(aq) \rightarrow Sn^{4+}(aq) + Mn^{2+}(aq)$

- Oxidation half reaction

  $Sn^{2+}(aq) \rightarrow Sn^{4+}(aq) + 2e^-$

- Reduction half reaction

  $MnO_4^-(aq) + 8H^+(aq) + 5e^- \rightarrow Mn^{2+}(aq) + 4H_2O(l)$

- Overall reaction at acidic condition

$$5Sn^{2+}(aq) \rightarrow 5Sn^{4+}(aq) + 10e^-$$
$$+\ \ 2MnO_4^-(aq) + 16H^+(aq) + 10e^- \rightarrow 2Mn^{2+}(aq) + 8H_2O(l)$$
$$\overline{5Sn^{2+}(aq) + 2MnO_4^-(aq) + 16H^+(aq) \rightarrow 5Sn^{4+}(aq) + 2Mn^{2+}(aq) + 8H_2O(l)}$$

- Overall reaction at basic condition

$$5Sn^{2+}(aq) \rightarrow 5Sn^{4+}(aq) + 10e^-$$
$$+\ \ MnO_4^-(aq) + 8H^+(aq) + 5e^- \rightarrow Mn^{2+}(aq) + 4H_2O(l)$$
$$\overline{5Sn^{2+}(aq) + 2MnO_4^-(aq) + 16H^+(aq) + 16OH^-(aq) \rightarrow 5Sn^{4+}(aq) + 2Mn^{2+}(aq) + 8H_2O(l) + 16OH^-(aq)}$$
$$\Rightarrow 5Sn^{2+}(aq) + 2MnO_4^-(aq) + 8H_2O(l) \rightarrow 5Sn^{4+}(aq) + 2Mn^{2+}(aq) + 16OH^-(aq)$$

13.2   Electrochemical Cell

13.2.1 Spontaneous electrochemical cell

Spontaneous electrochemical cell		Non-spontaneous electrochemical cell	
($+$) electrode : Cathode	Reduction	($+$) electrode : Anode	Oxidation
($-$) electrode : Anode	Oxidation	($-$) electrode : Cathode	Reduction
Thermodynamics	$\Delta G^\circ < 0$ spontaneous	Thermodynamics	$\Delta G^\circ > 0$ non-spontaneous

- **Galvanic cell**

($-$) electrode	: $Zn(s) \rightarrow Zn^{2+}(aq) + 2e^-$
($+$) electrode	: $Cu^{2+}(aq) + 2e- \rightarrow Cu(s)$
Overall reaction	: $Zn(s) + Cu^{2+}(aq) \rightarrow Zn^{2+}(aq) + Cu(s)$

<The role of salt bridge>
- Salt bridge if filled with the salt, for example, KNO_3
- It connect the two half cells without mixing with each other.
- The cation of salt moves to cathode, and the anion of salt moves to anode. theses flows of ions can maintain the electric neutrality of the solution

• **Standard reduction potential**

Standard reduction potential, E° is the reduction half-reactions with all solutes at standard states. Standard state means the 25℃, 1atm, and 1M of solution. The standard reduction potential is intensive properties which is independent from the number of electrons transferred in electrochemical cell

$$Ag^+(aq) + e^- \rightarrow Ag(s) \quad E° = + 0.80V$$
$$Cu^{2+}(aq) + 2e^- \rightarrow Cu(s) \quad E° = + 0.34V$$
$$2H^+(aq) + 2e^- \rightarrow H_2(g) \quad E° = \quad 0.00V$$
$$Zn^{2+}(aq) + 2e^- \rightarrow Zn(s) \quad E° = - 0.76V$$

$$Zn(s) + Cu^{2+}(aq) \rightarrow Zn^{2+}(aq) + Cu(s) \quad E° = +1.10V$$
$$Zn(s) + 2Ag^+(aq) \rightarrow Zn^{2+}(aq) + 2Ag(s) \quad E° = +1.56V$$
$$Cu(s) + 2Ag^+(aq) \rightarrow Cu^{2+}(aq) + 2Ag(s) \quad E° = +0.46V$$

13.2.2 Non-spontaneous electrochemical cell

• **Electrolytic cell : Electrolysis**

Electrolytic cell non-spontaneous cell. In order for electrolysis to occur, the power supply must be connected.

CuCl₂(aq) electrolysis		CuCl₂(l) electrolysis	
(+) electrode : Anode	$2Cl^-(aq) \rightarrow Cl_2(g) + 2e^-$	(+) electrode : Anode	$2Cl^-(l) \rightarrow Cl_2(g) + 2e^-$
(−) electrode : Cathode	$Cu^{2+}(aq) + 2e^- \rightarrow Cu(s)$	(−) electrode : Cathode	$Cu^{2+}(l) + 2e^- \rightarrow Cu(s)$
$2Cl^-(aq) + Cu^{2+}(aq) \rightarrow Cu(s) + Cl_2(g)$		$2Cl^-(l) + Cu^{2+}(l) \rightarrow Cu(s) + Cl_2(g)$	

NaCl(aq) electrolysis		NaCl(l) electrolysis	
(+) electrode : Anode	$2Cl^-(aq) \rightarrow Cl_2(g) + 2e^-$	(+) electrode : Anode	$2Cl^-(l) \rightarrow Cl_2(g) + 2e^-$
(−) electrode : Cathode	$2H_2O(l) + 2e^- \rightarrow H_2(g) + 2OH^-(aq)$	(−) electrode : Cathode	$Na^+(l) + e^- \rightarrow Na(s)$
$2Cl^-(aq) + 2H_2O(l) \rightarrow H_2(g) + Cl_2(g) + 2OH^-(aq)$		$2Cl^-(l) + 2Na^+(l) \rightarrow Cl_2(g) + 2Na(s)$	

Electrolysis of Water (Limited amount of salt must be added : Na₂SO₄, NaOH)	
(+) electrode : Anode	$2H_2O(l) \rightarrow O_2(g) + 4H^+(aq) + 4e^-$
(−) electrode : Cathode	$2H_2O(l) + 2e^- \rightarrow H_2(g) + 2OH^-(aq)$
$2H_2O(l) \rightarrow 2H_2(g) + O_2(g)$	

• **Electrolytic cell : Electroplating**

In electroplating, the plating material is oxidized at the anode and the material to be plated is plated at the cathode.

When iron is plated with silver

$(-)$ electrode   : $Ag+(aq)+e^- \rightarrow Ag(s)$

$(+)$ electrode   : $Ag(s) \rightarrow Ag+(aq)+e^-$

• **Faraday's Law**

$1A = 1C/s$ $1F = 96500C/mol\ e^-$

The electric current to electroplating the Fe(s) using the 1L of 0.1M of $CuSO_4(aq)$ in 1hr

$(0.1mol/L) \times (1L)Cu \times (2mol\ e^-/1mol\ Cu) \times (96500C/mol\ e^-)/3600s = 5.36C/s = 5.36A$

Chapter Review Questions   13. Electrochemistry

[241-242] refer to the following aqueous solutions

(A) Oxidation

(B) Reducing agent

(C) Cathode

(D) Anode

(E) Standard reduction potential

241 Negative electrode in electrolytic cell

242 The substance oxidized in redox reaction

243 Zinc(Zn) metal will reduce Cu^{2+}(aq) in solution

Because

Zinc(Zn) is a more active metal than copper

244 Electrolytic cell needs the input of the electric energy

Because

The reaction of electrolysis is nonspontaneous

245 The below is standard reduction potential (E^{o}_{red})for some substance.

$Na^+(aq)+e^- \rightarrow Na(s)$	$E^{o}_{red} = -2.71V$
$O_2(g)+4H^+(aq)+4e^- \rightarrow 2H_2O(l)$	$E^{o}_{red} = +1.23V$
$2H_2O(l)+2e^- \rightarrow H_2(g)+2OH^-(aq)$	$E^{o}_{red} = -0.83V$
$Cl_2(g)+2e^- \rightarrow 2Cl^-(aq)$	$E^{o}_{red} = +1.36V$

Which of the following is correct for the electrolysis of NaCl(aq) and NaCl(l)?

I. The products of cathode from electrolysis of NaCl(aq) and NaCl(l) are different.

II. Na(s) can not be obtained from electrolysis of NaCl(aq)

III. The change of mass of anode from electrolysis of NaCl(aq) and NaCl(l) are different.

(A) I only   (B) II only   (C) I, and II only

(D) II, and III only   (E) I, II, and III

246 the below is unbalanced redox reaction equation.

$..H^+(aq) + ...Fe(s) + ...NO_3^-(aq) \rightarrow ...Fe^{3+}(aq) + ...NO(aq) + ...H_2O(l)$

standard reduction potential

$NO_3^-(aq)/NO(aq) : +0.960V$ $\qquad\qquad$ $Fe^{3+}(aq)/Fe(s) : -0.44V$

Which of the following is correct?

I. The sum of coefficients as whole number is 10
II. The cell potential is $+1.40V$ and nonspontaneous reaction.
III. The mass of cathode decrease but that of anode remain constant.

(A) I only $\qquad\qquad$ (B) II only $\qquad\qquad$ (C) I, and II only

(D) II, and III only $\qquad\qquad$ (E) I, II, and IIII

247 Which of the following is correct for the unbalanced redox reaction represented below?

$...Pb(OH)_4^-(aq) + ...ClO^-(aq) \rightarrow ...PbO_2(s) + ...Cl^-(aq) + ...H_2O(l) + ...OH^-(aq)$

I. $Pb(OH)_4^-$ is oxidizing agent.
II. Oxidation number of Cl decreases from $+1$ to -1
III. The sum of coefficient as whole number is 11

(A) I only $\qquad\qquad$ (B) II only $\qquad\qquad$ (C) III only

(D) I, and II only $\qquad\qquad$ (E) II, and III only

248 Which of the following is correct for the reaction below

$...HMnO_4(aq) + ...H_2SO_3(aq) \rightarrow ...MnSO_4(aq) + ...H_2SO_4(aq) + ...H_2O(l)$

I. The sum of all coefficient as whole number is 15
II. $HMnO_4(aq)$ is oxidizing agent
III. 5 moles of electrons are involved per 1 mole of $HMnO_4$

(A) I only $\qquad\qquad$ (B) II only $\qquad\qquad$ (C) I, and II only

(D) II, and III only $\qquad\qquad$ (E) I, II, and III

249 Which of the following is correct for the redox reaction?

I. Na(sodium) is a stronger reducing agent than K(potassium) in gas phase.
II. Standard reduction potential(E^o_{red}) of $H^+(aq)$ is 0
III. In aqueous solution, Zn(zinc) is a stronger reducing agent than Cu(copper) because the standard reduction potential(E^o_{red}) of $Zn^{2+}(aq)$ is smaller than that of $Cu^{2+}(aq)$

(A) I only   (B) II only   (C) I, and II only

(D) II, and III only   (E) I, II, and III

250 $H_2O(l)$ can be decomposed into $H_2(g)$ and $O_2(g)$. Which of the following is correct Except for the reaction?

$$Reaction\ 1 : 2H_2O(l) \rightarrow O_2(g) + 4H^+(aq) + 4e^-$$
$$Reaction\ 2 : 2H_2O(l) + 2e^- \rightarrow H_2(g) + 2OH^-(aq)$$

(A) Reaction 1 is oxidation reaction

(B) Reaction 2 is cathode reaction

(C) 2F(Faraday) of electrons are needed for 1mole of $H_2(g)$

(D) Volume ratio of $H_2(g) : O_2(g)$ produced in decomposition of water is 1 : 1

(E) The standard cell potential is negative

251
$$...Cu(s) + ...NO_3^-(aq) + ...H^+(aq) \rightarrow ...Cu^{2+}(aq) + ...NO_2(g) + ...H_2O(l)$$

Which of the following is not correct for the this reaction ?

(A) The reaction is oxidation and reduction reaction

(B) Cu(s) is oxidized

(C) $H^+(aq)$ is reduced

(D) Entropy of the reaction increases

(E) Oxidation number of nitrogen is changed from $+5$ to $+4$

252 When the following reaction is balanced, what is the sum of coefficient as whole number?

$$...H^+(aq) + ...MnO_4^-(aq) + ...Fe^{2+}(aq) \rightarrow ...Mn^{2+}(aq) + ...Fe^{3+}(aq) + ...H_2O(l)$$

(A) 21   (B) 22   (C) 23   (D) 24   (E) 25

253 Which of the following is not correct for the redox reaction?

(A) reaction between metal and metallic salt

(B) reaction between metal and acid

(C) reaction between alkali metal and water

(D) reaction between transition metal and ligand

(E) reaction between elements to form a compound

254

$$...Cu^{2+}(aq) + ... I^-(aq) \rightarrow ...CuI(s) + ...I_2(s)$$

When the equation above is balanced and all coefficients are reduced to lowest whole-number terms, the sum of coefficients of the above reaction?

(A) 6   (B) 7   (C) 8   (D) 9   (E) 10

255 Which of the following compound is largest oxidation state in carbon?

(A) CO   (B) CH_4 (C) $CaCO_3$ (D) C_2H_5OH (E) HCOOH

256 Which of the following is not correct for the name and oxidation state of Cl(chlorine)?

(A) ClO^- hypochlorate $+1$
(B) ClO_2^- chlorite $+3$
(C) ClO_3^- chlorate $+5$
(D) ClO_4^- perchlorate $+7$
(E) Cl^- chloride -1

257 Which of the following is not correct for the Galvanic cell?

(A) Oxidation occurs at the anode
(B) Reduction occurs at (+) electrode
(C) Cathode is (+) electrode
(D) The reaction is not spontaneous
(E) E^o_{Cell} is positive

258 Which of the following is the equipment used in Galvanic cell to maintain the charge neutrality?

(A) Cathode   (B) Anode   (C) Voltmeter   (D) Salt bridge   (E) Bulb

259 Which of the following is not correct for the electrochemical data to set up the electrochemical cell?

$Ag+(aq)$	e^-	$\rightarrow Ag(s)$	$E^o = +0.80V$
$Cu^{2+}(aq)$	$2e^-$	$\rightarrow Cu(s)$	$E^o = +0.34V$

(A) Ag(s) is cathode
(B) Cu(s) is negative electrode
(C) E^o_{Cell} is 0.12V
(D) The mass of (+) electrode increases
(E) Cu(s) is oxidized

260 Al(s) can be obtained from electrolysis of Al_2O_3. Which of the following is not correct for the below reaction?

$$(-) \text{ electrode} : Al^{3+} + 3e^- \rightarrow Al$$
$$(+) \text{ electrode} : 6O^{2-} + 3C \rightarrow 3CO_2 + 12e^-$$

(A) $(-)$ electrode is cathode

(B) The reaction of $(+)$ electrode is oxidation

(C) Overall reaction is $4Al^{3+} + 6O^{2-} + 3C \rightarrow 4Al + 3CO_2$

(D) 12mole of electrons are needed for 1mole of Al

(E) The reaction is nonspontaneous at standard state because the standard cell potential is negative

14 Organic Chemistry

14.1 Organic Chemistry

14.1.1 Classification of organic compounds

Classification of organic compounds				Type of reaction
Saturated	Aliphatic	Alkane	C_nH_{2n+2}	Substitution
		Cycloalkane	C_nH_{2n}	Substitution
Unsaturated		Alkene	C_nH_{2n}	Addition
		Alkyne	C_nH_{2n-2}	Addition
		Cycloalkene	C_nH_{2n-2}	Addition
	Aromatic	Benzene	C_6H_6	Substitution

• **Nomenclature of Alkane**

Nomenclature of Alkane and properties of ten normal alkane					
Name	Formula	Molar mass	Melting point(℃)	Boiling point(℃)	structural isomer
Methane	CH_4	16	-182	-162	1
Ethane	C_2H_6	30	-183	-89	1
Propane	C_3H_8	44	-187	-42	1
Butane	C_4H_{10}	58	-138	0	2
Pentane	C_5H_{12}	72	-130	36	3
Hexnane	C_6H_{14}	86	-95	68	5
Heptane	C_7H_{16}	100	-91	98	9
Octane	C_8H_{18}	114	-57	126	18
Nonane	C_9H_{20}	128	-54	151	35

14.1.2 Isomer of organic compounds

Structural isomer (constructional isomer)	Stereoisomer	
	Geometrical isomer	Optical isomer(Enantiomer)
Atoms are arranged in a completely different way → Chain isomerism	Cis-Trans isomerism → Alkene	Isomers have opposite effects on plane polarized light (mirror image) → Sugar, Amino acid

14.2 Hydrocarbon Derivatives

Hydrocarbon derivative	Formula(condensed structural formula)	Example
Halohydrocarbons	$R-X$	CH_3I
Alcohols	$R-OH$	CH_3OH
Ethers	$R-O-R'$	CH_3OCH_3
Aldehydes	$R(H)-CHO$	$HCHO$
Ketones	$R-CO-R'$	CH_3COCH_3
Carboxylic acids	$R-COOH$	CH_3COOH
Ester	$R-COOR'$	CH_3COOCH_3

Chapter Review Questions | **14. Organic Chemistry**

[261-263] Refer to the following organic compounds

(A) Alcohol

(B) Aldehyde

(C) Ketone

(D) Carboxylic acid

(E) Amino acid

---

261 Alcohol can react with this compound to form ester

262 Compounds that can be isomer of aldehyde in same number of carbon

263 Contains both acidic and basic functional group

[264-270] Refer to the following organic compounds

(A) CH_3CH_2OH

(B) CH_2CH_2

(C) CH_3NH_2

(D) CH_3COOH

(E) CH_3OCH_3

---

264 Neutral and the most polar molecule is

265 Acidic polar molecule is

266 Basic polar molecule is

267 This compound is structural isomer of ethanol

268 This compound can react with carboxylic acid to form ester

269 This compound has both sp^2 and sp^3 carbons

270 This unsaturated hydrocarbon is not soluble in water

15 Nuclear Chemistry

15.1 Nuclear Stability and decay reaction

15.1.1 Nuclear stability

- All nuclides with 84 or more protons are unstable with respect to radioactive decay.
- Light nuclides are stable when Z(proton number) equals N(neutron number), that is, when the neutron-to-proton ratio is 1. However, for heavier elements the neutron-to-proton ratio required for stability is greater than 1 and increases with Z.

15.1.2 Radio decay reaction

Radio decay reaction	Reaction
beta(β) decay	$^{238}_{92}U \rightarrow {}^{238}_{93}Np + {}^{0}_{-1}e$
alpha(α) decay	$^{239}_{94}Pu \rightarrow {}^{235}_{92}U + {}^{4}_{2}He$
gamma(γ) decay	$^{234m}_{91}Pa \rightarrow {}^{234}_{91}Pa + {}^{0}_{0}\gamma$
positron emission	$^{18}_{9}F \rightarrow {}^{18}_{8}O + {}^{0}_{1}e$
electron capture	$^{103}_{46}Pd + {}^{0}_{-1}e \rightarrow {}^{103}_{45}Rh$

Radio decay reaction is first order reaction so half-life is constant and this independent on the temperature

Chapter Review Questions 15. Nuclear Chemistry

271 If a radioactive element has a half-life of 20 years, after 100 years, 20% of the initial amount remains.

Because

For each half-life, the remaining amount is reduced by half.

272 The energy of α particle is lager(more penetrate) than that of β particle

Because

The α particle is heavier than the β particle

273 α particle deflected to positive electrode when it enter the magnetic field

Because

Alpha particle is nucleus of He(helium)

274 $^{238}_{92}U \rightarrow ^{234}_{90}Th + ^{4}_{2}He$ is the transmutation decay of $^{238}_{92}U$

Because

Half-life of nuclear decay is constant and not affected by temperature

275 Which nuclear decay does the atomic number increase?

(A) alpha emission   (B) beta emission   (C) electron capture

(D) gamma decay   (E) positron emission

276 A radioactive isotope decay from 50g to 12.5g in 50days. What is the half-life of this radioactive isotope?

(A) 6.25day   (B) 12.5day   (C) 25.0day

(D) 37.5day   (E) 50.0day

277 Which of the following equipment is used to detect the radioactivity?

(A) Volumetric flask   (B) Calorimeter   (C) Geiger counter

(D) Pipet   (E) Funnel

278 Which of the following is correct for the blank?

$$_{27}^{57}\text{Co} + \boxed{} \rightarrow {}_{26}^{57}\text{Fe} + {}_{0}^{0}\gamma$$

(A) electron   (B) neutron   (C) positron

(D) α particle   (E) γ ray

279 Which of the following is correct for the blank?

$$_{86}^{222}\text{Rn} \rightarrow {}_{86}^{222}\text{Rn} + \boxed{}$$

(A) positron emission   (B) gamma ray emission   (C) alpha decay

(D) beta decay   (E) neutron emission

280 Which of the following is correct for the blank?

$$_{38}^{90}\text{Sr} \rightarrow {}_{-1}^{0}\beta + \boxed{}$$

(A) $_{37}^{87}\text{Rb}$ (B) $_{39}^{90}\text{Y}$ (C) $_{37}^{90}\text{Rb}$ (D) $_{40}^{90}\text{Zr}$ (E) $_{40}^{91}\text{Zr}$

MEMO

Final Full Test

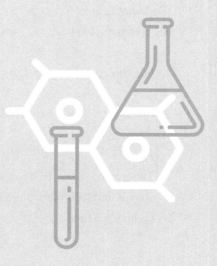

Chemistry Test **No.1**

Note : For all questions involving solution, assume that the solvent is water unless otherwise stated. Through the test the following symbols have the definitions specified unless otherwise noted.

H	=	enthalpy	atm	=	atmosphere(s)
M	=	molar	g	=	gram(s)
n	=	number of moles	J	=	joule(s)
P	=	pressure	kJ	=	kilojoule(s)
R	=	molar gas constant	L	=	liter(s)
S	=	entropy	mL	=	milliliter(s)
T	=	temperature	mm	=	millimeter(s)
V	=	volume	mol	=	mole(s)
V	=	volt(s)			

Part A

Directions : Each set of lettered choices below refers to the numbered statements or questions immediately following it. Select the one lettered choice that best fits each statement of answers each question and then fill in the corresponding circle on the answer sheet. A choice may be used once, more than once, or not at all in each set.

Questions 1-4 refer to the following types of bond.

(A) Polar covalent bond
(B) Non polar covalent bond
(C) Ionic bond
(D) Metallic bond
(E) Hydrogen bond

1. Intramolecular force in CO_2

2. Chemical bond between cation and free electron(s)

3. The reason why HF has a small molecular weight, but its boiling point is higher than other hydrogen halides.

Questions 4-7 refer to the following elements

(A) Ne
(B) O
(C) F
(D) Ca
(E) Na

4. has the most stable electron configuration at ground state.

5. has the largest second ionization energy

6. is the strongest oxidizing agent

7. has the largest radius when it is the same electron configuration as noble gas

Questions 8-11 refer to the following compounds

(A) $CuNO_3$

(B) Na_2CO_3

(C) $Pb(NO_3)_2$

(D) $ZnSO_4$

(E) Ca_3PO_4

8. reacts with HCl(aq) to form a gas denser than air

9. reacts with NaI(aq) to form a white precipitate

10. react with Na_2S to form a black precipitate

11. Its color is orange in frame reaction

Questions 12-15 refer to the following reactions.

(A) $CuSO_4 \cdot 10H_2O(s) \rightarrow CuSO_4(s) + 10H_2O(l)$

(B) $3HNO_2(g) \rightarrow HNO_3(g) + 2NO(g) + H_2O(g)$

(C) $Al(OH)_3(s) + NaOH(aq) \rightarrow Na[Al(OH)_4](aq)$

(D) $MgI_2(aq) + 2NaOH(aq) \rightarrow Mg(OH)_2(s) + 2NaI(aq)$

(E) $CaF_2(s) + HCl(aq) \rightarrow CaCl_2(aq) + 2HF(aq)$

12. is Lewis Acid and Base reaction but not Brønsted-Lowy Acid and Base

13. is oxidation–reduction reaction

14. is precipitation reaction

15. is dehydration reaction

Questions 16-18 refer to the following electron configurations of neutral atoms

(A) $1s^2 2s^2 2p^6$

(B) $1s^2 2s^2 2p^6 3s^2 3p^2$

(C) $1s^2 2s^2 2p^6 3s^2 3p^6 4s^2$

(D) $1s^2 2s^2 2p^6 3s^2 3p^6 4s^1 3d^5$

(E) $1s^2 2s^2 2p^6 3s^2 3p^6 4s^2 3d^5 4p^6 5s^2 4d^{10} 5p^6 6s^2 4f^3$

16. The second ionization energy is related to the d orbital electron

17. The electron configuration of lanthanide

18. The electron configuration of law material of semi–conductor

Questions 19-21 refer to the following aqueous solutions at 25℃

(A) 0.01M of NH_3(aq), $K_b(NH_3) = 1.8 \times 10^{-5}$

(B) 0.01M of $Ca(OH)_2$(aq)

(C) 0.01M of HCl(aq)

(D) 0.01M of $NaC_2H_3O_2$(aq), $K_a(HC_2H_3O_2) = 1.8 \times 10^{-5}$

(E) 0.01M of $(NH_4)_2SO_4$(aq)

19. $[H^+] < 0.01M$ and acidic

20. $[H^+] < 10^{-7}M$ and the weakest basic

21. When it reacts with HCl(aq), neutral salt is formed

Questions 22-23 **refer to the following pieces of laboratory equipment**

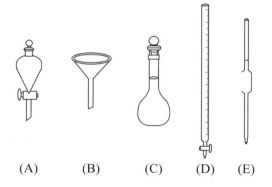

(A)   (B)   (C)   (D)   (E)

22. is used to take 25mL from 2M NaOH stock solution to make 0.1M NaOH(aq)

23. is used to separate the components applying the solubility difference in solvents.

Questions 24-25 **refer to the following**

(A) Polymerization
(B) Titration
(C) Vaporization
(D) Fractional crystallization
(E) Hydrolysis

24. Nylon 66, synthetic fiber, is synthesized from condensation reaction of hexamethylenediamine and adipic acid.

25. is used for refining substances from the mixture through solubility difference according to the temperature

PLEASE GO TO THE SPECIAL SECTION AT THE LOWER LEFT-HAND CORNER OF THE PAGE OF THE ANSWER SHEET YOU ARE WORKING ON AND ANSWER QUESTIONS 101-115 ACCORDING TO THE FOLLOWING DIRECTION.

Part B

Direction: Each question below consists of two statements. I in the left-hand column and II in the right-hand

column. For each question, determine whether statement I is true of false and whether statement II is true or false and fill in the corresponding T or F circles on your answer sheet. Fill in circle CE only if statement II is a correct explanation of the true statement I

EXAMPLE:

	I		II
EX1.	HNO_3 is a strong acid	Because	HNO_3 contains nitrogen.
EX2.	An atom of fluorine is electrically neutral.	Because	an fluorine contains an equal number of protons and electrons.

SAMPLE ANSWER

	I	II	CE*
EX1	● Ⓕ	● Ⓕ	○
EX2	● Ⓕ	● Ⓕ	●

	I		II
101.	^{12}C, ^{13}C is allotrope	Because	These atoms have the same physical and chemical properties
102.	The hydrogen ion concentration of a solution with a $pH = 2$ is 100 times that of a solution with a $pH = 4$.	Because	pH is the decimal logarithm of the reciprocal of the molarity of hydrogen
103.	$H_2O(l)$ is amphiprotic substance	Because	$H_2O(l)$ can donate or accepter the proton, H^+
104.	C_3H_8 and C_4H_{10} are the homologous series.	Because	The empirical formulas of the two compounds are the same.
105.	Generally, atomic number increases, atomic radius increases	Because	The repulsive force between electrons increases as the number of outermost electrons increases.

106. When 1g of Na_2CO_3 (molar mass 106g/mol) is added to 1g of NaCl(molar mass 58.5g/mol) and mixed, the mass % of Na(sodium) per unit mass decreases **Because** The mass % of Na(sodium) is smaller in NaCl than Na_2CO_3

107. The boiling point of decane($C_{10}H_{22}$) is lower than that of water(H_2O) **Because** There is only London dispersion force in $C_{10}H_{22}$, but only hydrogen bond in H_2O

108. The wavelength of line spectrum from n =2 to n=1 is larger than that of from n =4 to n=2 **Because** The energy difference is large when an electron transition with a large difference in the principal quantum number(n) occurs.

109. At a constant pressure, a certain amount of the volume of ideal gas increases by 1/273 of the 0℃ volume when the temperature increases by 1℃ **Because** Under these conditions, the volume of gas is proportional to the temperature in degrees Celsius

110. The energy released when $F^-(g)$ is formed by adding electrons to the neutral gaseous atom, F(g), is called ionization energy. **Because** The electron of the $F^-(g)$ is the same as stable electron configuration of noble gas

111. The $C-O$ bond length in carbonate ion, CO_3^{2-}, are all the same. **Because** The geometry of carbonate ion, CO_3^{2-}, is trigonal planar

$$NO_2^-(aq) + H_2O(l) \rightleftharpoons HNO_2(aq) + OH^-(aq)$$

112. NO_2^- is Lewis base **Because** $NO_2^-(aq)$ accepts the proton, H^+

113. When the temperature of the aqueous solution decreases when dissolved in water of the ionic solid, the solubility can be increased by increasing the temperature. **Because** This dissolution reaction is an endothermic reaction.

114. The boiling point of 10%wt glucose(molar mass 180g/mol) solution is higher than 20%wt sucrose(molar mass 342g/mol) solution **Because** The number glucose molecules in nuit mass of water is larger than that of sucrose

115. The volume of ideal gas is zero at 0K **Because** The volume of gas molecule decreases as the temperature decreases

Part C

Direction: Each of the questions or incomplete statements below is followed by five suggested answer or completions. Select the one that is best in each case and then fill in the corresponding circle on the answer sheet.

26. $...Al_2Si_2O_5(OH)_4 + ...H^+ \rightarrow ...SiO_2 + ...Al^{3+} + ...H_2O$

Which of the following is correct for the reaction represented above?

I. The reaction is oxidation–reduction reaction
II. The molecular structure of SiO_2 is the same as CO_2
III. The sum of coefficients as whole number is 16

(A) I only   (B) II only   (C) III only
(D) I, II only   (E) I, II, and III

27. Which of the following is correct for the Na, Mg, O, and F?

(A) Atomic radius $O < F < Mg < Na$
(B) Isoelectronic radius $Mg^{2+} < Na^+ < F^- < O^{2-}$
(C) (first) ionization energy $Mg < Na < O < F$
(D) Electronetativity $Na < Mg < F < O$
(E) Second ionization energy $Mg < Na < F < O$

28. In gaseous reaction, 15g of A(g) reacts with 8g of B(g) completely and produce the gas C(g) When initially 10L of A and 5L of B exist, and after the reaction, there is only 10L of C(g). Which of the following is correct for the reaction?

I. 2mol of B is required for 1mol of A to react completely

II. Molar mass ratio $A : B = 15 : 4$
III. The ratio of density of gas $A : B : C = 15 : 16 : 23$ at constant temperature and pressure

(A) I only   (B) II only   (C) III only
(D) I, II only   (E) II, III only

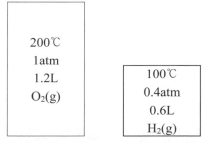

29. Which of the following is correct for the $O_2(g)$ and $H_2(g)$ in rigid containers?

(A) The density of $O_2(g)$ is smaller than $H_2(g)$
(B) The average kinetic energy of $O_2(g)$ is 2 times larger than that of $H_2(g)$
(C) The average speed of $O_2(g)$ molecule is smaller than $H_2(g)$
(D) When two gas are mixed in 4.0L of container the partial pressure of $O_2(g)$ is 2.5 times larger than that of $H_2(g)$
(E) The kinetic energy of any $O_2(g)$ molecule is always greater than that of $H_2(g)$

30. Which of the following is not correct?

(A) It is difficult to obtain the alkali metals as elements because of their large reactivity in nature.

(B) Generally, metal oxide is base anhydride

(C) Al_2O_3 can react both acid and base

(D) Generally, halogens are good reducing agent in chemical reaction

(E) Transition elements partially filled with electrons in d orbital are colored

31. What is the mass of NaOH(s)(molar mass 40.0g/mol) required to make 500mL of 0.4M NaOH(aq) aqueous solution?

(A) 8.0g   (B) 12.0g   (C) 16.0g

(D) 20.0g   (E) 24.0g

32. $N_2O_4(g) \rightleftharpoons 2NO_2(g)$

	$N_2O_4(g)$	$NO_2(g)$
$\Delta H_f^o(kJ\cdot mol^{-1})$	10.0	34.0
$S^o(J\cdot mol^{-1}\cdot K^{-1})$	304	241

initially, there is only $N_2O_4(g)$ in evacuated container. the total pressure of container remains the same in reaction. Which of the following is correct for the reaction represented above?

(A) The forward reaction is exothermic

(B) When increasing the temperature, equilibrium constant decrease

(C) The forward reaction can be spontaneous at low temperature

(D) The sum of bond energy of reactants is larger than that of products

(E) The equilibrium constant of forward reaction is affected by pressure of container

33. $...Fe^{2+}(aq)+...Cr_2O_7^{2-}(aq)+...14H^+(aq) \rightarrow$
$6Fe^{3+}(aq)+..Cr^{n+}(aq)+...H_2O(l)$

Which of the following is correct for the sum of coefficient as whole number and value of n?

	sum of coefficient	n
(A)	34	1
(B)	34	2
(C)	34	3
(D)	36	2
(E)	36	3

34. $...C_{18}H_{36}O_2(s)+...O_2(g) \rightarrow$
$...CO_2(g)+...H_2O(g)$

Which of the following is correct for the reaction represented above?
(N_A : Avogadro's number, 6.02×10^{23})

(A) 20mol of $O_2(g)$ is required for complete combustion of 1mol of $C_{18}H_{36}O_2(s)$

(B) The volume of $CO_2(g)$ produced from complete combustion of 1mol of $C_{18}H_{36}O_2(s)$ is $16\times22.4L$ at 0℃, 1atm

(C) When 1mol of $C_{18}H_{36}O_2(s)$ reacts with 13mol of $O_2(g)$ completely, the volume of $CO_2(g)$ produced is $9\times22.4L$ at 0℃, 1atm

(D) When 0.5mol of $C_{18}H_{36}O_2(s)$ is reacted with excess of $O_2(g)$ the total volume of produced gas is $10\times22.4L$

(E) When 2mol of $C_{18}H_{36}O_2(s)$ is reacted with excess of $O_2(g)$ the total number of produced gas is $36\times N_A$

35. which of the following is suitable for the molecular formula of hydrocarbon composed of 82% C and 18% H as mass%?

(A) C_2H_4 (B) C_3H_8 (C) C_4H_6
(D) C_4H_{10} (E) C_5H_{12}

36. When 0.05M of $FeCl_3$(aq) 200mL is mixed with 0.04M of $FeCl_2$(aq) 300mL. What is the molarity of Cl^-(aq)?

(A) 0.044M   (B) 0.064M   (C) 0.076M
(D) 0.108M   (E) 0.216M

37. 300mL of He gas at 27℃, 400mmHg is heated to 127℃ at constant pressure. And pressure is added to the gas the final pressure of gas is 600mmHg but the volume of gas is no more changed. What is the final temperature?

(A) 127℃   (B) 227℃   (C) 327℃
(D) 427℃   (E) 527℃

38. 10mL of HCl (aq) with pH = 2 was taken and diluted with distilled water to make a 1L of solution. What is the pH of diluted solution?

(A) 3   (B) 4   (C) 5
(D) 6   (E) 7

39. How much heat energy is required to change the 100.0g of ethanol(C_2H_5OH) from the solid state with freezing point of $-180℃$ to gas state with boiling temperature, 78℃?

Specific heat capacity of C_2H_5OH(s) : 0.98(J/g℃)
Specific heat capacity of C_2H_5OH(l) : 2.30(J/g℃)
Heat of fusion of C_2H_5OH(s) : 220J/g
Heat of valorization of C_2H_5OH(l) : 860J/g

(A) 81kJ   (B) 126kJ   (C) 167kJ
(D) 183kJ   (E) 215kJ

$N_2(g) + O_2(g) \rightarrow 2NO(g)$	$\Delta H^o_{rxn} = 183kJ$
$1/2N_2(g) + O_2(g) \rightarrow NO_2(g)$	$\Delta H^o_{rxn} = 33kJ$

40. Which of the following is correct enthalpy change for the reaction represented below?

$2NO(g) + O_2(g) \rightarrow 2NO_2(g)$

(A) $-117kJ$ (B) $-150kJ$ (C) 117kJ
(D) 150kJ   (E) 216kJ

HPO_4^{2-}(aq) $+ H_2O$(l) $\rightarrow H_2PO_4^-$(aq) $+ OH^-$(aq)

41. Which of the following is correct for the HPO_4^{2-} in the reaction represented above?

(A) Lewis acid
(B) Brønsted−Lowry acid
(C) Brønsted−Lowry base
(D) Oxidizing agent
(E) Reducing agent

42. Which of the following reactions is expected to increase in entropy when forward reaction occurs?

(A) $Na^+(g) + Cl^-(g) \rightarrow NaCl(s)$
(B) $CH_4(g) + 2O_2(g) \rightarrow CO_2(g) + 2H_2O(l)$
(C) $4NH_3(g) + 5O_2(g) \rightarrow 4NO(g) + 6H_2O(g)$
(D) $4Fe(s) + 3O_2(g) \rightarrow 2Fe_2O_3(s)$
(E) $4Fe_2S(s) + 11O_2(g) \rightarrow 2Fe_2O_3(s) + 8SO_2(g)$

43. What is the density of a 16.0g of spherical substance with $64cm^3$?

(A) $0.3.g/cm^3$ (B) $0.30.g/cm^3$
(C) $0.25.g/cm^3$ (D) $0.250.g/cm^3$
(E) $0.2500.g/cm^3$

44. Which of the following is correct for the concept of chemical equilibrium?

(A) At equilibrium, the concentration of reactants are the same as that of products
(B) At equilibrium, the concentration of reactants are larger than that of products
(C) At equilibrium, the concentration of reactants are smaller than that of products
(D) At equilibrium, the rate of forward reaction is the same as that of reverse reaction
(E) At equilibrium, the rates of forward reaction and reverse reaction are the zero

45. Which of the following is expected to have the highest pH after the reaction?

(A) 1L of 0.1M $HC_2H_3O_2$ and 1L of 0.1M NaOH
(B) 2L of 0.5M $HC_2H_3O_2$ and 1L of 0.1M NaOH

(C) 1L of 0.1M NH_3 and 1L of 0.1M HCl
(D) 2L of 0.1M NH_3 and 1L of 0.2M HCl
(E) 1L of 0.1M $HC_2H_3O_2$ and 1L of 0.1M NH_3

$$2NO(g) + 2H_2(g) \rightarrow N_2(g) + 2H_2O(g)$$

46. The table below is the experimental results of measuring the reaction rate represented above according to the initial concentration of the reactants.

Ex.	[NO], M	[H], M	initial rate (M/s)
1	0.2	0.1	8.00×10^{-4}
2	0.4	0.1	3.20×10^{-3}
3	0.1	0.2	4.00×10^{-4}

Which of the following is correct for the reaction?

(A) Forward reaction is exothermic because entropy decreases when a forward reaction occurs.
(B) The rate of NO(g) is consumed is 2 times faster than the rate of $N_2(g)$ formed
(C) As the temperature increases, the reaction rate may increase or decrease.
(D) The unit of rate constant is $M^2L^{-1}s^{-1}$
(E) The rate law is $k[NO][H_2]^2$ (k : rate constant)

$$NH_4CO_2NH_2(s) \rightleftharpoons 2NH_3(g) + CO_2(g)$$
$$\Delta H_{rxn}^\circ = +160kJ$$

47. Which of the following is not correct for the reaction represented above?

(A) At hight temperature the reaction can be spontaneous
(B) As the temperature increases, the partial pressure of $CO_2(g)$ increases at equilibrium

(C) Periodic removal of $NH_3(g)$ can increase the amount of product from the same amount of reactants

(D) Adding $NH_4CO_2NH_2(s)$ to the equilibrium, the reaction shift to right

(E) Adding pressure to the equilibrium, the reaction shift to left

$$CaF_2(s) \rightleftharpoons Ca^{2+}(aq) + 2F^-(aq)$$
$$K_{sp} = 4.00 \times 10^{-11}(25℃)$$

48. Which of the following is not correct for the dissolution of $CaF_2(s)$? $(K_a(HF) = 1.8 \times 10^{-4})$

(A) The solubility of $CaF_2(s)$ in $NaF(aq)$ is smaller than in pure water, $H_2O(l)$

(B) The solubility of $CaF_2(s)$ is affected by the pH of solution

(C) The solubility product, $K_{sp} = [Ca^{2+}][F^-]^2$

(D) When $Ca(NO_3)_2(s)$ is added to equilibrium the solubility decreases

(E) The solubility $CaCO_3(s)$ is larger than $CaF_2(s)$ $(K_{sp}(CaCO_3) = 8.7 \times 10^{-9})$

49. The pH of 0.1M of weak acid, HA is 3. What is the acid dissociation constant, $K_a(HA)$?

(A) 10^{-3} (B) 10^{-4} (C) 10^{-5}
(D) 10^{-6} (E) 10^{-7}

50. The empirical formula CH_2O, and the molar mass is 60g/mol which of the following is correct for the molecular formula?

(A) CH_2O (B) $C_2H_4O_2$ (C) $C_3H_6O_3$
(D) $C_4H_8O_4$ (E) $C_5H_{10}O_5$

51. Which of the following does not change the value by increasing the amount?

(A) length   (B) mass   (C) temperature
(E) volume   (C) energy

$$...CaC_2(s) + ...H_2O(l) \rightarrow ...Ca(OH)_2(aq) + ...C_2H_2(g)$$

52. 100g of $CaC_2(s)$ (molar mass 64g/mol) and 100g of $H_2O(l)$ were reacted. When the limiting reactant was consumed completely, which of the following is not correct?

(A) The sum of coefficients as whole number is 5

(B) The limiting reactant is $CaC_2(s)$

(C) The volume of produced gas is $\frac{100}{64} \times 2 \times 22.4$L at 0℃, 1atm

(D) The mass of produced gas is $\frac{13}{32} \times 100$g

(E) The reaction is not oxidation-reduction reaction

53. Which of the following is correct for the cell potential of a galvanic cell using the reaction below?

$Fe^{3+}(aq) + e^- \rightarrow Fe^{2+}(aq)$	$E_{red}^o = 0.771\,V$
$Cu^{2+}(aq) + 2e^- \rightarrow Cu(s)$	$E_{red}^o = 0.340\,V$

(A) $0.091\,V$ (B) $0.431\,V$ (C) $-0.431\,V$
(D) $1.202\,V$ (E) $-1.202\,V$

54. $...C_3H_6O(l) + ...O_2(g) \rightarrow ...CO_2(g) + ...H_2O(l)$

Which of the following is correct for the reaction represented above?

standard enthalpy of formation

$C_3H_6O(l) : -250kJ/mol, \ CO_2(g) : -394kJ/mol$
$H_2O(l) : -286kJ/mol$

I. When forward reaction occurs, the entropy of system increase.
II. When 1mol of $C_3H_6O(l)$ react with $O_2(g)$ completely 1790kJ of heat energy is released
III. The sum of coefficient as whole number is 10

(A) I only (B) II only (III) III only
(D) I, II only (E) II, III only

55. Which of the following is correct for the reaction rate?

(A) Endothermic reaction is slower than exothermic reaction
(B) Catalyst can not change the heat of reaction
(C) The equilibrium constant of fast reaction is larger than slow reaction
(D) As temperature increases, the rate of exothermic reaction decrease
(E) As temperature decreases, activation energy increases

56. Which of the following is correct for the electron configuration of ground sate in period 3 nonmetal atom?

(A) $1s^22s^22p^53s^1$
(B) $1s^22s^22p^63s^3$
(C) $1s^22s^22p^63s^23p^1$
(D) $1s^22s^22p^63s^23p^4$
(E) $1s^22s^22p^53s^23p^64s^23d^6$

57. The reaction order of radioisotope, I-131 decay reaction is first order. The half life of I-131 is 8 days. which of the following is correct for the time needed 10.00g of I-131 to decay and 1.25g of I-131 remains?

(A) 8days (B) 12days (C) 16days
(D) 20days (E) 24days

$N_2(g) + 2O_2(g) \rightarrow N_2O_4(g) \quad \Delta H_1$
$2N_2O(g) + 3O_2(g) \rightarrow 2N_2O_4(g) \quad \Delta H_2$

58. Which of the following is correct for the enthalpy change(ΔH_3) of the reaction represented below?

$2N_2(g) + O_2(g) \rightarrow 2N_2O(g) \quad \Delta H_3$

(A) $\Delta H_1 - 2\Delta H_2$ (B) $2\Delta H_1 - \Delta H_2$
(C) $-2\Delta H_1 + \Delta H_2$ (D) $4\Delta H_1 - 2\Delta H_2$
(E) $-4\Delta H_1 + 2\Delta H_2$

59. How many grams of $Ca(OH)_2$ (molar mass : 74.0g/mol) are needed to make 40.0mL of 0.100M solution?

(A) 0.148g (B) 0.296g (C) 0.444g
(D) 0.592g (E) 0.740g

$A(g) + 3B(g) \rightleftharpoons 2C(g) \quad Kc$

60. When 4.0mol of A(g) and 5mol of B(g) are added in 1.00L of evacuate rigid container. When the reaction reaches equilibrium. the mole fraction of B(g) is $\frac{2}{7}$. Which of the following is correct for the equilibrium constant(Kc)?

(A) 1   (B) 3   (C) 5

(D) $\frac{1}{2}$ (E) $\frac{1}{6}$

61. Which of the following is correct for the name of compound?

(A) Mg_2N_3 diamagnesium trinitride

(B) Hg_2SO_4 mercury(I) sulfate

(C) NO_2 mononitrogen dioxide

(D) $AlCl_3$ aluminium trichloride

(E) $Pb(NO)_2$ lead(II) dinitrate

62. Which of the following mixture has the highest temperature through the neutralization with 30mL of 0.04M of HCl(aq)?(molar mass of NaOH : 40.0g/mol)

(A) 10mL of 0.02M NaOH(aq)

(B) 30mL of 0.04M NaOH(aq)

(C) 20mL of 0.05M NaOH(aq)

(D) 40mg of NaOH(s)

(E) 60mg of NaOH(s)

63. Which of the following is correct for the energy profile represented below?

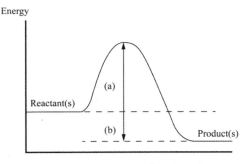

Reaction pathway

(A) When temperature increases, the fraction of reactants more than Ea(activation energy) decreases

(B) When the concentration of reactants, the rate constant increases

(C) When the catalyst is used, the reaction mechanism is changed.

(D) The change of rate of reaction is related with the change of (b)

(E) The catalyst is consumed as reactant or product

64. Which of the following substance can increase the solubility of $Cr(OH)_3(s)$?

$$Cr(OH)_3(s) \rightleftharpoons Cr^{3+}(aq) + 3OH^-(aq) \quad Ksp$$

Which of the following is correct for the reaction represented above?(temperature : 25℃)

(A) When NaOH added to equilibrium, the reaction shifts to right

(B) When pH is decreased the solubility product (Ksp) increases

(C) At equilibrium, pH is affected by the mass of $Cr(OH)_3(s)$ remained.

(D) At equilibrium, the concentration of $Cr^{3+}(aq)$ is the same as $OH^-(aq)$

(E) When $Cr(NO_3)_3(s)$ is added to equilibrium, the solubility of $Cr(OH)_3(s)$ is decreases

65. Which of the following has the highest oxidation number of sulfur(S) or nitrogen(N)?

(A) N_2O_5 (B) NO_3^- (C) NH_3

(D) H_2SO_4 (E) $SOCl_2$

66. 50.0mL of 0.3M of H_2SO_4(aq) is needed to titrate the 40mL NaOH(aq) that contains a droplet of phenolphthalein at 25℃. which of the following is correct?

(A) The pH at endpoint is less than 7
(B) The color of phenolphthalein change from colorless to pink
(C) Initial pH of NaOH(aq) is larger than 13
(D) Initial concentration of NaOH(aq) is 0.5M
(E) Na^+(aq) and SO_4^{2-}(aq) are the net ions

67. When 40.0mL of 0.30M of $CaCl_2$(aq) is added to 120mL of 0.10M $AgNO_3$(aq). which of the following is correct for the reaction?

(A) Yellow precipitate forms
(B) Limiting reactant is $CaCl_2$(aq)
(C) Net ions are Ca^{2+}(aq) and NO_3^-(aq)
(D) After the reaction, the concentration $[NO_3^-]$ is 0.075M
(E) After the reaction, the total concentration of ions is 0.15M

68. Which of the following is not correct for safety and rule in chemistry laboratory?

(A) The remaining reagents should not be returned to the bottle again
(B) Do not apply force to the glass tube through a cork or rubber stopper without proper lubrication
(C) Always wear eye protection glasses with proven stability during the experiment
(D) The chemicals used can be treated with the individual judgment of the experimenter
(E) Laboratories are available on time and with administrator permission

69. Which of the following is the least needed in a fractional distillation experiment?

(A) crucible
(B) distilling flask
(C) condenser
(D) hot plate
(E) thermometer

70. How many electrons are needed to get 1 mol of Al from the Al_2O_3 molten?

(A) 1mol   (B) 2mol   (C) 3mol
(D) 4mol   (E) 6mol

Chemistry Test No.2

Note : For all questions involving solution, assume that the solvent is water unless otherwise stated. Through the test the following symbols have the definitions specified unless otherwise noted.

H	=	enthalpy	atm	=	atmosphere(s)
M	=	molar	g	=	gram(s)
n	=	number of moles	J	=	joule(s)
P	=	pressure	kJ	=	kilojoule(s)
R	=	molar gas constant	L	=	liter(s)
S	=	entropy	mL	=	milliliter(s)
T	=	temperature	mm	=	millimeter(s)
V	=	volume	mol	=	mole(s)
V	=	volt(s)			

Part A

Directions : Each set of lettered choices below refers to the numbered statements or questions immediately following it. Select the one lettered choice that best fits each statement of answers each question and then fill in the corresponding circle on the answer sheet. A choice may be used once, more than once, or not at all in each set.

Questions 1-3 Refer to the following substances.

(A) Al_2O_3
(B) NaCl
(C) C(s, graphite)
(D) CCl_4
(E) SiO_2(s, glass)

1. is the amphoteric oxide. (A)

2. is the amorphorous solid. (E)

3. is the network solid with electric conductivity.(C)

Questions 4-6 Refer to the following.

(A) First ionization energy
(B) Electron affinity
(C) Bond dissociation energy
(D) Heat of vaporization
(E) Activation energy

4. is the amount of energy required to transform the substance from liquid to gas phase. (D)

5. is the minimum energy required to remove the valence electron from the gaseous ground state neutral atom. (A)

6. is the energy required to breakdown the gaseous covalent bond.(C)

Questions 7-9 Refer to the following substances.

(A) H_2O_2
(B) CO_2
(C) OF_2
(D) N_2O_5
(E) MgO

7. is the compound with highest oxidation state oxygen atom

8. is the molecule with three-dimensional structure

9. is the substance that causes acid rain.

Questions 10-13 Refer to the neutral atoms with following electron configurations

(A) $1s^2$
(B) $1s^22s^22p^3$
(C) $1s^22s^22p^5$
(D) $1s^22s^12p^63s^1$
(E) $1s^22s^22p^63s^23p^64s^13d^9$

10. is strong oxidizing agent in chemical reaction.

11. is the excited state of noble gas.

12. has the highest ionization energy.

13. has various oxidation state and its $+2$ ion shows color.

Questions 14-16 Refer to the following molecules.

(A) H_2S
(B) SO_3
(C) CCl_3F
(D) NH_3
(E) HCN

14. is the nonpolar molecule with polar covalent bond

15. is the polar molecule with triple bond

16. is the pollutant that destroys the stratospheric ozone layer

Questions 17-19 Refer to the following elements.

(A) Tin(Sn)
(B) Silicon(Si)
(C) Carbon(C)
(D) Copper(Cu)
(E) Magnesium(Mg)

17. can be used for cathodic protection to prevent corrosion of Fe(iron)

18. is the raw material of semiconductor.

19. is the component of brass

Questions 20-22 **Refer to the followings**

(A) Alpha particle
(B) Beta particle
(C) Gamma ray
(D) Positron
(E) Nucleus

20. is release when U-238 decay to Th-234

21. was discovered form Rutherford's gold foil experiment

22. has the shortest wavelength in light

Questions 23-25 **Refer to the following solutions.**

(A) 0.4M, 1L of Na_2CO_3(aq) is mixed with 0.2M, 1L of $Ba(OH)_2$(aq)

(B) 0.2M, 1L of CH_3COOH(aq) is mixed with 0.2M, 1L of $Ca(OH)_2$(aq)

(C) 0.1M, 1L of NaCl(aq) is mixed with 0.1M, 1L of KNO_3(aq)

(D) 0.1M, 1L of H_2SO_4(aq) is mixed with 0.2M 1L of NaOH(aq)

(E) 0.1M, 1L of H_2SO_4(aq) is mixed with 0.2M 1L of NH_3(aq)

23. is the most acidic solution

24. can neutralize 0.2M, 1L of HCl(aq)

25. can produce white precipitate

PLEASE GO TO THE SPECIAL SECTION AT THE LOWER LEFT-HAND CORNER OF THE PAGE OF THE ANSWER SHEET YOU ARE WORKING ON AND ANSWER QUESTIONS 101-115 ACCORDING TO THE FOLLOWING DIRECTION.

Part B

Direction: Each question below consists of two statements. I in the left-hand column and II in the right-hand
column. For each question, determine whether statement I is true of false and whether statement II is true or false and fill in the corresponding T or F circles on your answer sheet. Fill in circle CE only if statement II is a correct explanation of the true statement I

EXAMPLE:

	I		II
EX1.	HNO$_3$ is a strong acid	Because	HNO$_3$ contains nitrogen.
EX2.	An atom of fluorine is electrically neutral.	Because	an fluorine contains an equal number of protons and electrons.

SAMPLE ANSWER

	I	II	CE*
EX1	● Ⓕ	● Ⓕ	○
EX2	● Ⓕ	● Ⓕ	●

	I		II
101.	NH$_3$(aq) can be act as Lewis base in aqueous solution	Because	The NH$_3$(aq) molecule forms hydrogen bond with H$_2$O(l) molecule
102.	Benzene(C$_6$H$_6$) is unsaturated hydrocarbon	Because	All of the carbon atoms in benzene are sp^3 hybridization
103.	The boiling point of H$_2$O(l) can be higher than 100℃	Because	Boiling occurs when the vapor pressure and the external pressure are the same.
104.	O$_3$ is nonpolar molecule	Because	All of the molecules have no net dipole
105.	C(s, diamond) is molecule	Because	The tetrahedral unit of diamond is the same as CH$_4$(methane)
105.	The diffusion rate of N$_2$(g) is higher than that of O$_2$(g) at same temperature	Because	The atomic radius of N(nitrogen) is smaller than that of O(oxygen)

107. $H_2(g)$ from the reaction between copper, Cu(s) and hydrochloric acid, HCl(aq) can be collected through water displacement

Because

The solubility of $H_2(g)$ in water increase as the partial pressure of $H_2(g)$ above the water at constant temperature

108. The pH of 0.1M $Ca(OH)_2$(aq) is higher than 13 at 25℃

Because

The concentration of hydrogen ion, H^+(aq) is lower than 10^{-13}M when 0.1M of $Ca(OH)_2$ is fully ionized in water at 25℃

109. When the temperature increase in endothermic equilibrium reaction $N_2O_4(g) \rightleftharpoons 2NO_2(g)$, the equilibrium constant increase

Because

Increasing the temperature in the endothermic reaction slows the reverse reaction rate.

110. Vapor pressure of ethanol, CH_3CH_2OH(l) is higher than that of hexane, C_6H_{14}(l) at normal boiling point

Because

Ethanol molecules can form hydrogen bonds between molecules, but hexane can not.

111. The radius of Na^+(g) is smaller than Na(g)

Because

The outermost electron is attracted in positively charged ion more strongly than in its neutral atom

112. When copper, Cu(s) is added to $AgNO_3$(aq), total ion concentration decrease as reaction proceeds

Because

Cu(s) reacts with Ag^+(aq) in a mol ratio 2 : 1

113. The boiling point of HBr(l) is higher than that of HCl(l)

Because

Dipole-dipole force of HBr(l) is higher than that of HCl(l)

114. $1s^2 2s^2 2p^5 3s^1$ is the ground electron configuration of Na(sodium)

Because

The 3s electron is valence electron of neutral Na(sodium) atom

115. When the kelvin temperature of the gas in the rigid container is doubled, the pressure is doubled.

Because

The collision frequency of gas particles to the wall of container per unit time is doubled

Part C

Direction: Each of the questions or incomplete statements below is followed by five suggested answer or completions. Select the one that is best in each case and then fill in the corresponding circle on the answer sheet.

26. The below is the line spectrum of gaseous hydrogen atom

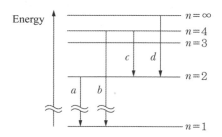

Which of the following is not correct for the graph?(the energy of hydrogen atom is $E_n = -\dfrac{13.6e\,V}{n^2}$)

(A) The wavelength of spectrum a is shortest.

(B) The line spectrum b is the wavelength range of UV(ultraviolet).

(C) The sum of energy $a+d$ is the same as the ionization energy of hydrogen atom at ground state.

(D) The energy difference between $n=1$ and $n=2$ larger than that of between $n=\infty$ and $n=2$

(E) Only one the line spectrum represented above is visible.

27. The below is the enthalpy change of some reaction

reaction 1 $4NO_2(g) + O_2(g) \rightarrow 2N_2O_5(s)$

$\Delta H = -219kJ$

reaction 2 $2NO(g) + O_2(g) \rightarrow 2NO_2(g)$

$\Delta H = -114kJ$

Calculate the ΔH for the reaction below using the data represented above?

$2N_2O_5(s) \rightarrow 4NO(g) + 3O_2(g)$

(A) 9kJ   (B) $-9kJ$ (C) 105kJ

(D) $-105kJ$ (E) $+447kJ$

28. The below is the unbalanced aqueous reaction.

$...Cu^{2+}(aq) + ...I^-(aq) + ...\boxed{}(aq)$ $\rightarrow 2CuI(s) + S_4O_6^{2-}(aq)$

Which of the following is not correct for the reaction represented above?

(A) $Cu^{2+}(aq)$ is reduced.

(B) $I^-(aq)$ is neither oxidized nor reduced, but net ion in the reaction.

(C) $S_2O_3^{2-}$ is possible for the blank

(D) The blue color of solution disappears as the reaction proceeds.

(E) The sum of reaction coefficient as whole number is 8.

29. Which of the following has the lowest freezing point?

(A) 0.1M of NaCl(aq)
(B) 0.2M of $C_{12}H_{22}O_{11}$(aq)
(C) 0.05M of Na_3PO_4(aq)
(D) 0.15M of $AgNO_3$(aq)
(E) 0.3M of $HC_2H_3O_2$(aq)

30. When 200mL of 0.1M Ca(OH)$_2$(aq) mixed with 300mL of 0.05M Na_2CO_3(aq), The concentration of Ca^{2+}(aq) is

(A) 0.01M   (B) 0.02M   (C) 0.03M
(D) 0.04M   (E) 0.05M

31. The volume of NaOH(aq) is 1.440L to neutralize the 300mL of 0.200M H_3PO_4(aq) completely. what is the molarity of NaOH(aq)

(A) 0.100M   (B) 0.125M   (C) 0.250M
(D) 0.375M   (E) 0.500M

32. ...$CaCO_3$(s) + ...$HC_2H_3O_2$(aq) →
    ...$Ca(C_2H_3O_2)_2$(aq) + ...H_2O(l) + ...CO_2(g)

When 0.2g piece of calcium carbonate(molar mass 100g/mol) reacts with 0.1M of excess of $HC_2H_3O_2$(aq), which of the following is not correct for the reaction at standard state(STP)?

(A) This is a type of neutralization reaction
(B) The volume of gas produced is 44.8mL
(C) When 0.1M of excess HCl(aq) is used instead of $HC_2H_3O_2$(aq), the volume of gas produced increase.

(D) Entropy increases when a forward reaction proceeds
(E) If the pieces of $CaCO_3$ is substituted for powder type of $CaCO_3$ with same mass, the gas production rate increase

33. ...MnO_4^-(aq) + ...Fe^{2+}(aq) + ...H^+(aq) →
    ...Mn^{2+}(aq) + ...Fe^{3+}(aq) + ...H_2O(l)

When 200mL of 0.100M $KMnO_4$(aq) is added to 800mL of $Fe(NO_3)_2$(aq) at acidic condition, the reaction is completed. Which of the following is not correct for the reaction?

(A) MnO_4^- is oxidizing agent
(B) MnO_4^- can be indicator to determine the endpoint.
(C) The sum of reaction coefficient as whole number is 24.
(D) The concentration of NO_3^- at endpoint is 0.200M
(E) The initial concentration of $Fe(NO_3)_2$(aq) is 0.100M

34. The below is solubility curve for some solute in water.

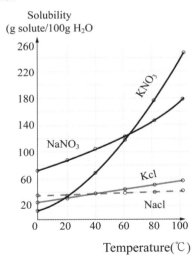

Which of the following is not correct for the solubility for represented above?

(A) The solubility depends on the species of solute.

(B) At 20℃ The solubility of $NaNO_3$ is highest of the solute represented above.

(C) When KCl is dissolved in water, the temperature of the solution decreases.

(D) When 50g of 80℃ saturated $KNO_3(aq)$ is cooled to 60℃, 40g of $KNO_3(aq)$ is crystallized.

(E) All points on each curve are saturated solutions.

35. 2mol of $N_2O_4(g)$ and 1mol of CO(g) are injected in 2L of rigid container. The initial total pressure is 6atm. $N_2O_4(g)$ and CO(g) react and NO(g), $CO_2(g)$, and $NO_2(g)$ are produced. The limiting reactant is consumed completely and temperature remains constant. Which of the following is correct?

(A) The limiting reactant is $N_2O_4(g)$

(B) The total pressure of container decrease as forward reaction proceeds.

(C) When the reaction is completed, partial pressure of each gas remained is not the same.

(D) When the reaction is completed, the partial pressure of $CO_2(g)$ is 2atm.

(E) $N_2O_4(g)$ is reducing agent in forward reaction.

36. Which of the following is not correct for the element with ground electron configuration, $1s^2 2s^2 2p^6 3s^2 3p^6 4s^1 3d^{10}$?

(A) Its atomic number is 29

(B) Generally it is oxidized in chemical reactions.

(C) It may have some oxidation states in compounds.

(D) +2 ion have color

(E) The electron configuration of +1 ion is $[Ar]4s^1 3d^9$

37. Which of the following is the right comparison for some periodicity

(A) Atomic radius   Na $<$ Mg

(B) Ionic radius $O^{2-} <$ Al^{3+}

(C) (First) ionization energy   F $<$ Na

(D) Electronegativity   N $<$ O

(E) Second ionization energy   Na $<$ Ne

38. Heating the hydrate can evaporate the water. After sufficiently heating 100g of $CuSO_4 \cdot xH_2O$, ($CuSO_4$ molar mass : 160g/mol) the mass of remain is 64.0g. What is the x?

(A) 1   (B) 2   (C) 3

(D) 4   (E) 5

39. $Ca_3(PO_4)_2(s) \rightleftharpoons 3Ca^{2+}(aq) + 2PO_4^{3-}(aq)$
$$Ksp = 2.1 \times 10^{-33}$$

Which of the following is correct for the reaction represented above?

(A) At equilibrium, $3[Ca^{2+}] = 2[PO_4^{3-}]$

(B) The expression of $Ksp = [Ca^{2+}]^2 / [PO_4^{3-}]^3$

(C) The solubility increases in $Ca(OH)_2(aq)$ solution.

(D) The solubility is about 10^{-7}M

(E) Solubility product(Ksp) decreases when $Na_3PO_4(s)$ is added to equilibrium.

40. $...Cu(s) + ...KCN(aq) + ...O_2(g) + ...H_2O(l)$
    $\rightarrow ...K[Cu(CN)_2](aq) + ...KOH(aq)$

Which of the following is correct for the sum of coefficient as whole number of the reaction represented above?

(A) 20   (B) 21   (C) 22

(D) 23   (E) 24

41. At constant temperature, when 3L of 4atm NO(g) and 5L of $O_2(g)$ are mixed in a 4L rigid container, the total pressure is 6atm. Which of the following is correct?

I. The initial pressure of $O_2(g)$ before mixing is 2atm.

II The partial pressure of NO(g) is the same as $O_2(g)$ in mixture.

III. The average speed of NO(g) is larger than that of $O_2(g)$

(A) I only   (B) II only

(C) I and II only   (D) II and III only

(E) I, II, and III

42. $...C_6H_{14}(l) + ...O_2(g) \rightarrow ...CO_2(g) + ...H_2O(l)$

Which of the following is correct?

I. If the reaction proceeds in the rigid container, the total pressure increases.

II. When 43g of $C_6H_{14}(l)$ consumed completely, the volume of $CO_2(g)$ gas produced is 67.2L at 0℃, 1atm.

III. The mass of $O_2(g)$ reacted is smaller than that of $CO_2(g)$ produced.

(A) I only   (B) II only

(C) I and II only   (D) II and III only

(E) I, II, and III

43. The below is the heating curve of solid substance at constant pressure, 1atm. The heat energy added to the system is constant.

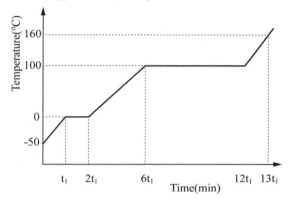

Which of the following is not correct for the graph?

(A) The solid substance is pure substance.

(B) The specific heat capacity of solid is half of liquid.

(C) The heat of vaporization is 6 times larger than that of heat of fusion.

(D) The vapor pressure of the liquid substance is 1atm at 100℃

(E) Temperature and average kinetic energy is remain the same during the phase transition.

44. Which of the following is not oxidation－reduction reaction?

(A) $2Fe_2O_3(s) \rightarrow 4Fe(s) + 3O_2(g)$

(B) $2NaI(aq) + Cl_2(aq) \rightarrow 2NaCl(aq) + I_2(aq)$

(C) $2Mg(s) + CO_2(g) \rightarrow 2MgO(s) + C(s)$

(D) $MnO_2(s) + 4HCl(aq) \rightarrow MnCl_2(aq) + 2H_2O(l) + Cl_2(g)$

(E) $2PCl_5(g) \rightarrow PCl_4^+(g) + PCl_6^-(g)$

45. $CO(g) + H_2O(g) \rightleftharpoons CO_2(g) + H_2(g) + energy$

Initially, there are only $CO(g)$ and $H_2O(g)$ in rigid container. Which of the following is not correct for the reaction?

(A) At equilibrium, partial pressure of $CO_2(g)$ and $H_2(g)$ are the same.

(B) As increasing the temperature, the equilibrium constant is decreasing.

(C) When $He(g)$ is added to equilibrium state, there is no shift of the reaction.

(D) When $CO(g)$ is added to equilibrium state, The equilibrium constant increases.

(E) The total pressure is always the same regardless of progress of the reaction.

46. When 30mL of 0.2M of $NaNO_3(aq)$ is mixed with 50mL 0.6M of $Ca(NO_3)_2(aq)$. which of the following is correct for the concentration of $NO_3^-(aq)$ in mixutre?

(A) 0.125M   (B) 0.250M   (C) 0.450M

(D) 0.825M   (E) 0.925M

47. Which of the following is not correct for the comparison of ideal gas and real gas?

(A) Ideal gas and real gas have the mass.

(B) The phase of ideal gas does not change regardless of pressure applied, but real gas does.

(C) Ideal gas and real gas have the volume of gas particles.

(D) There is no intermolecular force in ideal gas but is in real gas

(E) The behavior of real gas is similar with ideal gas at high temperature and low pressure.

48. The boiling point of $0.1m$ $C_6H_{12}O_6(aq)$ solution is $0.052\,°C$ higher than normal boiling point of $H_2O(l)$ at 1atm. Which of the following is not correct for the boiling point of each solution?

(A) 0.1m of $HC_2H_3O_2(aq)$ $100.104\,°C$

(B) 0.05m of $NaCl(aq)$ $100.052\,°C$.

(C) 0.1m of $CaCl_2(aq)$ $100.156\,°C$

(D) 0.2m of $CH_3OH(aq)$ $100.104\,°C$

(E) 0.1m of $NaOH(aq)$ $100.104\,°C$

49. Which of the following is correct for the C_2H_2 and C_4H_8

(A) Hybrid orbital of all carbon(C) atoms in C_4H_8 are the same.

(B) The mass % of carbon(C) in C_2H_2 is smaller than that in C_4H_8

(C) The number of sigma bond in C_2H_2 is 4.

(D) All the carbon(C) atoms exist in the same plane in C_4H_8

(E) The boiling point of C_2H_2 is lower than that of C_4H_8 at constant pressure.

50. How much $Na_2SO_4(s)$(molar mass : 142g/mol) is needed to prepares the 200 mL of 0.1M $Na_2SO_4(aq)$?

(A) 1.42g   (B) 2.84g   (C) 14.2g

(D) 28.4g   (E) 42.6g

51. When 10mL of 2.0M of $Ca(OH)_2$ was taken and diluted with distilled water, the concentration of $[OH^-]$ is 0.1M. Which of the following is correct for the distilled water to dilute the concentrate solution?(Assume volumes are additive)

(A) 190mL   (B) 200mL   (C) 290mL

(D) 300mL   (E) 390mL

52. $2ClO_2(aq) + 2OH^-(aq) \rightarrow ClO_3^-(aq) + ClO_2^-(aq) + H_2O(l)$

The below is the concentration of reactants and initial rate of the reaction represented above.

Experiment	$[ClO_2]$ (M)	$[OH^-]$ (M)	initial rate (M/s)
1	0.01	0.05	3.2×10^{-2}
2	0.02	0.05	1.28×10^{-1}
3	0.005	0.2	3.2×10^{-2}

Which of the following is correct for the reaction?

(A) The reaction is first order for ClO_2

(B) When $[ClO_2] = 0.0025M$, $[OH^-] = 0.1M$, the initial rate is $4.0 \times 10^{-3}M/s$

(C) The unit of rate constant, is $M^{-3}s^{-1}$

(D) OH^- is the catalyst

(E) The activation energy of forward reaction is larger than that of reverse reaction

53. $CH_3NH_2(g)$ is occupied 3.02×10^2mL at 38℃ with pressure of 0.250atm. At which of the temperature(K) is correct when it occupies 4.37×10^2mL with pressure of 500torr?

(A) $\dfrac{4.37 \times 10^2 \times 0.250}{311 \times \frac{500}{760} \times 3.02 \times 10^2}(K)$

(B) $\dfrac{4.37 \times 10^{-1} \times \frac{500}{760}}{311 \times \frac{250}{760} \times 3.02 \times 10^{-1}}(K)$

(C) $\dfrac{311 \times 4.37 \times 10^2 \times \frac{500}{760}}{0.250 \times 3.02 \times 10^2}(K)$

(D) $\dfrac{311 \times 4.37 \times 10^2 \times 500}{250 \times 3.02 \times 10^2}(K)$

(E) $\dfrac{4.37 \times 10^{-1} \times 500}{311 \times 250 \times 3.02 \times 10^{-1}}(K)$

54. A compound compose of Ca(calcium), C(carbon), H(hydrogen), and O(oxygen) is analyzed. The mass % are Ca 25%, C 30%, H 5%, and O 41%. which of the following is correct for the compound?

(A) $Ca(CH_3O)_2$ (B) $CaC_2H_3O_2$

(C) $Ca(C_2H_3O_2)_2$ (D) $Ca_2C_2H_3O_2$

(E) $Ca_2(CH_3O)_3$

55. A compound contains C(carbon) 13.6% as mass %. Which of the following another elements is contained in the compound when the molar mass is the lowest as possible?(Assume carbon atom satisfies with octet rule)

(A) H   (B) N   (C) O

(D) F   (E) Cl

56. Which of the following is correct for the name of ion?

(A) ClO_3^- chlorite

(B) MnO_4^- manganate

(C) SO_3^{2-} sulfite

(D) S^{2-} sulfate

(E) HCO_3^- carbonite

57. $Zn^{2+}(aq) + 4NH_3(aq) \rightleftharpoons [Zn(NH_3)_4]^{2+}(aq)$

Which of the following is correct for the expression of equilibrium constant, Kc?

(A) $[Zn^{2+}][NH_3]^4 - [Zn(NH_3)_4]^{2+}$

(B) $[Zn^{2+}][NH_3]^4[Zn(NH_3)_4]^{2+}$

(C) $\dfrac{[NH_3]^4[Zn(NH_3)_4^{2+}]}{[Zn^{2+}]}$

(D) $\dfrac{[Zn(NH_3)_4^{2+}]}{[Zn^{2+}][NH_3]^4}$

(E) $\dfrac{[Zn(NH_3)_4^{2+}]^4}{[Zn^{2+}]^4[NH_3]}$

58. The below is the distribution of speed of $N_2(g)$ molecules at specific temperature.

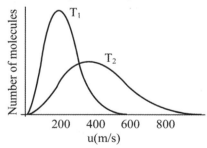

Which of the following is correct for the graph?

(A) Average kinetic energy is higher at T_1 than T_2

(B) All molecules have the same kinetic energy at T_1

(C) All molecules at T_2 are higher than T_1

(D) As temperature increase both average kinetic energy and average speed of molecule increase

(E) Intermolecular force between $N_2(g)$ is larger at T_2 than T_1

59. $C(s, graphite) + H_2O(g) \rightleftharpoons CO(g) + H_2(g)$

Standard enthalpy of formation $CO(g)$ is $-115kJ/mol$, and $H_2O(g)$ is $-242kJ/mol$. Which of the following is correct for the reaction represented above?

(A) The activation energy of forward reaction is smaller than reverse reaction

(B) When external pressure is added to equilibrium state and the volume of container decreases, the reaction shift to forward reaction.

(C) When noble gas is added to the equilibrium state, the reaction shift to reverse reaction

(D) The forward reaction can be spontaneous at high temperature.

(E) When C(s, graphite) added to the equilibrium state, the reaction shift to forward reaction

60. Which of the following is not a pair of conjugate acid and base?

	Conjugate acid	Conjugate base
(A)	H_2S	S^{2-}
(B)	HNO_2	NO_2^-
(C)	H_2O	OH^-
(D)	NH_3	NH_2^-
(E)	CH_4	CH_3^-

61. Which of the following is correct comparison for the melting and boiling point?

(A) melting point   NaCl >   MgO
(B) melting point SiO_2 > CO_2
(C) melting point   Na   >   Ca
(D) boiling point   HCl >   HBr
(E) boiling point N_2 > O_2

62. $2NO(g) + O_2(g) \rightleftharpoons 2NO_2(g)$ $Kp = 1.0 \times 10^8$

Initially, There are only $NO(g)$ and $O_2(g)$ in 1.00L of rigid container. Before the reaction partial pressure of $NO(g)$ is 2.0×10^{-2}atm and $O_2(g)$ is 1.0×10^{-2}atm Which of the following is correct for the reaction?

(A) The equilibrium constant as molarity, Kc, is the same as the Kp

(B) At equilibrium, the partial pressure of $NO(g)$ is 1.0×10^{-4}atm

(C) As increases the temperature, the equilibrium constant increases.

(D) The sum of bond enthalpy of products are larger than that of reactants

(E) When the forward reaction occurs, the entropy of system increases.

63. 400mL of $H_3PO_4(aq)$ is titrated with 600mL of 0.01M $Ca(OH)_2(aq)$ is consumed. Which of the following is correct for the reaction?

(A) Initial concentration of $H_3PO_4(aq)$ 0.01M

(B) The electrical conductivity at the neutralization point is highest during the titration

(C) The pH of mixture is lower than 7 at end point

(D) Before the titration the pH of $H_3PO_4(aq)$ is lower than 2

(E) The net ionic equation of neutralization is $H^+(aq) + OH^-(aq) \rightarrow H_2O(l)$

64. The below is the phase diagram of H_2O

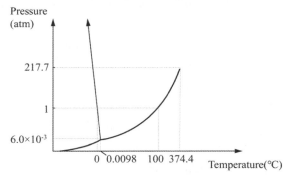

Which of the following is not correct?

(A) The density increases $H_2O(g) < H_2O(s) < H_2O(l)$

(B) The boiling point of $H_2O(l)$ increases as atmospheric pressure increases

(C) Above 374℃, even when pressure is applied to water vapor, condensation can not be observed.

(D) At 25℃, 1atm, $H_2O(l)$ is the most stable phase in three phases

(E) Because pressure of Triple point is much lower than 1atm, sublimation can not be observed in nature.

65. $...C_3H_3N_3O_3 + ...NO_2 \rightarrow ...N_2 + ...CO_2 + ...H_2O$

Which of the following is correct for the sum of coefficients as whole number?

(A) 81   (B) 82   (C) 83
(D) 84   (E) 85

66. $Cu(s) + 2AgNO_3(aq)(aq) \rightarrow Cu(NO_3)_2(aq) + 2Ag(s)$

which of the following is correct for the reaction?
(Atomic mass : Ag 107.87, Cu 63.55)

(A) $Cu^{2+}(aq)$ is stronger oxidizing agent than $Ag^+(aq)$

(B) The number of electrons transferred per 1mol of Ag(s) is twice that of Cu(s)

(C) When 1mol of Cu(s) react with $AgNO_3(aq)$ completely, the mass of the solution is reduced by 152.19g

(D) When 0.1mol of Cu(s) react completely, the total number of ions decreases by 0.2 mol.

(E) Cooper(Cu) electrode can be cathode, and silver(Ag) electrode can be anode in galvanic cell

67. When 3.70g of $Ca(OH)_2(s)$ (molar mass 74.0 g/mol) is added to 0.2M of 300mL $HNO_3(aq)$ with indicator phenolphthalein, which of the following is correct?

(A) Precipitate forms

(B) $NO_3^-(aq)$ is the largest number ion after the reaction

(C) The color of phenolphthalein change form pink to colorless

(D) The pH of the solution is greater than 5 and less than 7

(E) During the neutralization reaction, the electrical conductivity increases

68. Chromium(Cr) is used to make stainless alloys by mixing with iron. The density of chromium (Cr) is 7.19g/cm³. when 35.35g chromium is mixed with 295.0g of iron.

Which of the following is correct for the volume of chromium(Cr)?

(A) 2.10cm³ (B) 3.42cm³ (C) 4.92cm³
(D) 4.917cm³ (E) 5.00cm³

69. Which combination of quantum numbers is correct? (n : principal quantum number l : angular momentum quantum number, m_l : magnetic quantum number, m_s : spin quantum number)

(A) n=1 l=1 m_l=−2 m_s=+1/2
(B) n=2 l=0 m_l=−2 m_s=−1/2
(C) n=2 l=1 m_l=−1 m_s=+1/2
(D) n=3 l=2 m_l=−2 m_s=+1
(E) n=4 l=3 m_l=−2 m_s=−1

70. $...NaN_3(s) \rightarrow ...Na(s) + ...N_2(g)$

Sodium azide(NaN_3, molar mass 65g/mol) decomposition reactions are used in air bag. Initially, 0.1mol of $NaN_3(s)$ is in evacuated rigid 10L of container Which of the following is correct for the reaction?

(A) $N_2(g)$ is very reactive gas.
(B) The sum of coefficient as whole number is 8
(C) When $NaN_3(s)$ is completely decomposed, 8.4g of $N_2(g)$ is formed
(D) When $NaN_3(s)$ is completely decomposed, the pressure of $N_2(g)$ is $\frac{0.15 \times 0.0821 \times 298}{10}$ atm
(E) When % yield is 50%, the mass of Na(s) formed is 1.15g

Chemistry Test No.3

Note : For all questions involving solution, assume that the solvent is water unless otherwise stated. Through the test the following symbols have the definitions specified unless otherwise noted.

H	=	enthalpy	atm	=	atmosphere(s)
M	=	molar	g	=	gram(s)
n	=	number of moles	J	=	joule(s)
P	=	pressure	kJ	=	kilojoule(s)
R	=	molar gas constant	L	=	liter(s)
S	=	entropy	mL	=	milliliter(s)
T	=	temperature	mm	=	millimeter(s)
V	=	volume	mol	=	mole(s)
V	=	volt(s)			

Part A

Directions : Each set of lettered choices below refers to the numbered statements or questions immediately following it. Select the one lettered choice that best fits each statement of answers each question and then fill in the corresponding circle on the answer sheet. A choice may be used once, more than once, or not at all in each set.

Questions 1-4 refer to the following substances

(A) SO_3
(B) C(s, graphite)
(C) NaCl
(D) NH_3
(E) SiO_2(glass)

1. is pollutant from the combustion of fuel that can produce the acid rain

2. is network solid with electric conductivity

3. can form a hydrogen bond with each other

4. is amorphous solid that lacks a well defined arrangement

Questions 5-8 refer to the following molecules

(A) H_2S
(B) CF_2Cl_2
(C) SO_2
(D) CO_2
(E) HCN

5. is polar molecule with one pi bond

6. is polar molecule with linear shape

7. has two identical resonance structure

8. is nonpolar molecules with polar covalent bonds

Questions 9-13 refer to the following reactions

(A) A solution of 0.1M, 10mL $Ba(NO_3)_2$ and 0.1M 10mL Na_2SO_4 are mixed.
(B) Solid $CaCO_3$ is added to HCl solution.
(C) Solid sodium added to water.
(D) A solution of $AgNO_3$ and excess of ammonia mixed.
(E) A zinc solid added to a solution of $Cu(SO)_4$

9. A gaseous compound is produced

10. A complex ion is formed

11. A basic solution is formed

12. A double displacement reaction with no further reaction

13. A redox reaction with deposition of reddish metal

Questions 14-17 refer to the following elements

(A) sodium
(B) cobalt
(C) tin
(D) Aluminium
(E) silicon

14. is a component of bronze

15. is the most abundant metal in nature

16. is a component of semiconductor

17. is a element with partially filled 3d orbital

Questions 18-20 refer to the following definition

(A) Arrhenius acid
(B) Lewis base
(C) Brønsted-Lowry base
(D) Lewis acid
(E) Brønsted-Lowry acid

18. is the substance that produce hydrogen ion(s) in aqueous solution

19. is the substance that electron pair(s) donor in chemical reaction

20. central metal in complex ion is the example of this

Questions 21-25 refer to the following definition

(A) Standard reduction potential
(B) First ionization energy
(C) Electronegativity
(D) Electron affinity
(E) Standard enthalpy of formation

21. The energy required to remove one mole of electron from gaseous neutral atom

22. Heat change that results when 1 mole of the compound is formed from its elements at standard states

23. The ability of an atom to attract toward itself the electrons in a chemical bond

24. The energy change associated with the addition of an electron to a gaseous atom

25. is the tendency for a chemical species to be reduced in aqueous solution

PLEASE GO TO THE SPECIAL SECTION AT THE LOWER LEFT-HAND CORNER OF THE PAGE OF THE ANSWER SHEET YOU ARE WORKING ON AND ANSWER QUESTIONS 101-115 ACCORDING TO THE FOLLOWING DIRECTION.

Part B

Direction: Each question below consists of two statements. I in the left-hand column and II in the right-hand column. For each question, determine whether statement I is true of false and whether statement II is true or false and fill in the corresponding T or F circles on your answer sheet. Fill in circle CE only if statement II is a correct explanation of the true statement I

EXAMPLE:

	I		II
EX1.	HNO$_3$ is a strong acid	Because	HNO$_3$ contains nitrogen.
EX2.	An atom of fluorine is electrically neutral.	Because	an fluorine contains an equal number of protons and electrons.

SAMPLE ANSWER

	I	II	CE*
EX1	● Ⓕ	● Ⓕ	○
EX2	● Ⓕ	● Ⓕ	●

	I		II
101.	CH$_4$ is nonpolar molecules	Because	The molecular geometry of CH$_4$ is square planar
102.	The pH of 1.0×10^{-2}M hydrochloric acid solution is 2	Because	Hydrochloric acid is completely ionized in water
103.	There are 3mol of N$_2$(g) and 2mol of O$_2$(g) in rigid container with total pressure 1atm, partial pressure of N$_2$(g) is 0.3atm.	Because	The partial pressure of each gas component is proportional to the mole number(s) at the same temperature and same volume in gaseous mixture.
104.	In the conversion of NH$_3$ to NO$_2$, nitrogen is oxidized.	Because	The electronegativity of nitrogen is larger than hydrogen but smaller than oxygen.
105.	The first ionization energy of oxygen is smaller than that of fluorine at the ground state.	Because	There is(are) more electron pair(s) in 2p orbital in fluorine than oxygen.

106. To dilute the sulfuric acid, H_2SO_4, water added to concentrated sulfuric acid.   Because   The dilution of concentrated acid with water is exothermic.

107. Applying pressure to ice make it water.   Because   The density of ice is smaller than water

108. Catalyst can change the equilibrium position.   Because   Catalyst can lowering the activation energy in both forward and reverse reaction simultaneously

109. At the equilibrium, forward reaction and reverse reaction no longer occurs   Because   At the equilibrium, there is no change of concentration of reactants and products in concerned with macroscopic observation.

110. When 1mol of $C_3H_8(g)$ is reacted with 4mol of $O_2(g)$ completely, limiting reactant is $C_3H_8(g)$.   Because   Limiting reactant determine the theoretical yield of product.

111. The average speed of helium gas is twice that of methane gas at same temperature   Because   The average speed of ideal gas is inversely proportional to the molar mass at same temperature

112. The exothermic reaction is spontaneous.   Because   The potential energy of reactant(s) is(are) larger than that of product(s) in exothermic reaction.

113. The empirical formulas of C_2H_4 and C_6H_{12} are the same.   Because   The percent composition of each element are the same for same empirical formula.

114. The boiling point of 0.1-molal solution of sodium chloride is higher than 0.1-molal solution of glucose at same pressure.   Because   The boiling point of solution is affected by the total molal concentration of particles in solution regardless of the type of solution at same pressure.

115. The normal boiling point of H_2O is higher than H_2S Because   There are hydrogen bonds between H_2O molecules but are not in H_2S molecules.

Part C

Direction: Each of the questions or incomplete statements below is followed by five suggested answer or completions. Select the one that is best in each case and then fill in the corresponding circle on the answer sheet.

26. As the atomic number of group 1A metals is increasing,

(A) The ionization energy is increasing
(B) The atomic radius of is decreasing
(C) The melting point is decreasing
(D) The density is decreasing
(E) The electronegativity is increasing

27. Which of the following is different oxidation number of oxygen?

(A) Na_2O
(B) CO_2
(C) NO_2
(D) OF_2
(E) CuO

28. Which of the following is not correct for the Bohr's atomic model?

(A) Energy levels of electron are quantized
(B) When electron transits from lower energy level to higher level, discrete energy is absorbed
(C) The intensity of energy that atom absorbs or emits in transition of electron is inversely proportional to energy difference
(D) The wavelength of Lyman series are always shorter than Balmer series
(E) Energy difference between $n = 1$ and $n = 2$ is larger than between $n = 2$ and $n = 4$

29. Which of the following is not the characteristics of metals?

(A) Metals are good conductor of heat and electricity
(B) Generally metals are oxidizing agent in chemical reactions with nonmetals
(C) Metals can form a ionic compound with halides
(D) Generally metal oxides are basic in aqueous solution
(E) Metals are more abundant than nonmetals in the periodic table of elements

30. Which of the following elements has the largest number of unpaired electron in sub-shells in ground state?

(A) Na   (B) B
(C) Cu   (D) N
(E) O

31. Which of the following is correct explanation for the reason that transition metals shows color?

(A) They can form complex ions
(B) They can absorb the visible light
(C) They can release the ultraviolet light
(D) They can exist in various oxidation states
(E) They can form a ionic compound

32. Which of the following is not correct for the ground state of each element?

(A) N $1s^2 2s^2 2p^3$

(B) Ar $1s^2 2s^2 2p^6 3s^2 3p^6$

(C) Cr $1s^2 2s^2 2p^6 3s^2 3p^6 4s^1 3d^5$

(D) As $1s^2 2s^2 2p^6 3s^2 3p^6 4s^2 3d^{10} 4p^2$

(E) Sr $1s^2 2s^2 2p^6 3s^2 3p^6 4s^2 3d^{10} 4p^6 5s^2$

33. The below is the combustion reaction of $C_3H_8(g)$

$$C_3H_8(g) + 5O_2(g) \rightarrow 3CO_2(g) + 4H_2O(l)$$

The initial mole number of $C_3H_8(g)$ and $O_2(g)$ is 1mol and 10mol in each, and before the reaction, there are only reactants exist, which of the following is the not correct for the reaction?

(A) The limiting reactant is $C_3H_8(g)$

(B) If the reaction occurred in rigid container the total pressure of container decreases after the reaction

(C) If the reaction occurred at constant pressure, the volume of container increase after the reaction

(D) The total mole number of gases remained is 8mol

(E) If the percent yield 60% the mole number of $CO_2(g)$ produced is 1.8mol

34. Which of the following is not the result of chemical change?

(A) Gaseous product is evolved from the heating of sodium hydrogen carbonate($NaHCO_3$)

(B) When $Zn(s)$ is added to $CuSO_4(aq)$, the color of solution changes

(C) When sodium chloride($NaCl$) is dissolved, electric current is formed

(D) When $Na_2SO_4(aq)$ and $Ca(NO_3)_2(aq)$ are mixed, white precipitation is formed

(E) Excess ammonia is added to $ZnSO_4(aq)$, complex ion is formed

35. Which of the following is the correct expression of equilibrium constant, Kc, for the reaction below?

$$CaCO_3(s) \rightleftharpoons CaO(s) + CO_2(g)$$

(A) $Kc = \dfrac{[CaO][CO_2]}{[CaCO_3]}$

(B) $Kc = \dfrac{[CO_2]}{[CaCO_3]}$

(C) $Kc = \dfrac{[CaCO_3]}{[CaO][CO_2]}$

(D) $Kc = \dfrac{1}{[CO_2]}$

(E) $Kc = [CO_2]$

36. The below is the equilibrium equation.

$$2NO_2(g) \rightleftharpoons N_2O_4(g)$$

if the partial pressure of $NO_2(g)$ is 2 times of $N_2O_4(g)$ and total pressure is 6atm at equilibrium, which of the following is the equilibrium constant, Kp, of the reaction?

(A) 1/16   (B) 1/8

(C) 1/4   (D) 8

(E) 16

37. What is the mol ratio of Fe : O if the mass percent of Fe is 70% in a compound of iron and oxygen?(Atomic mass : Fe 55.8, O 16.00)

(A) 1 : 1
(B) 1 : 2
(C) 2 : 1
(D) 2 : 3
(E) 3 : 2

38. Which of the following has two unshared electron pairs of central atom in Lewis structure?

(A) NH_3
(B) H_3O^+
(C) NH_4^+
(D) OH^-
(E) OF_2

39. Which of the following is correct expression of the volume of gas when 25℃, 1.0atm, 30L gas is changed to 30℃ and 1.5 atm?

(A) $\dfrac{30}{25} \times \dfrac{1140}{760} \times 30L$

(B) $\dfrac{303}{298} \times \dfrac{1.5}{1.0} \times 30L$

(C) $\dfrac{25}{30} \times \dfrac{1.5}{1.0} \times 30L$

(D) $\dfrac{303}{298} \times \dfrac{760}{1140} \times 30L$

(E) $\dfrac{30}{25} \times \dfrac{1.5}{1.0} \times 30L$

40. When 1mol of Cu(s) react with 4mol of HNO_3(aq) completely, which of the following is possible one product of the reaction below?

$$Cu(s) + 4HNO_3(aq) \rightarrow Cu(NO_3)_2(aq) + \ldots + 2H_2O(l)$$

(A) NO(g)   (B) NO_2(g)
(C) N_2O(g)   (D) NO_3(g)
(E) N_2(g)

41. The below is the unbalanced oxidation-reduction reaction

$$Fe^{2+}(aq) + Cr_2O_7^{2-}(aq) + H^+(aq) \rightarrow Fe^{3+}(aq) + Cr^{3+}(aq) + H_2O(l)$$

Which of the following is correct for the reaction above?

(A) Fe^{2+}(aq) is oxidizing agent
(B) $Cr_2O_7^{2-}$(aq) is oxidized
(C) The mole number of electron transferred between $Cr_2O_7^{2-}$(aq) and Cr^{3+} is 6mol per 1mol of $Cr_2O_7^{2-}$(aq)
(D) The sum of reaction coefficient of reactants is smaller than that of products
(E) The oxidation number of Cr in $Cr_2O_7^{2-}$(aq) is $+7$

42. Which of the following is the most accurate in delivery of liquid?

(A) buret
(B) volumetric flask
(C) erlenmeyer flask
(D) beaker
(E) pipette

43. Which of the following 0.1M of solutions has the highest pH?

(A) HCl

(B) H_2SO_4

(C) Na_2CO_3

(D) $NaHCO_3$

(E) $HC_2H_3O_2$(ethanoic acid)

44. What volume of a 0.200molar solution of sulfuric acid is required to neutralize 50 milliliters of 0.800molar solution of sodium hydroxide completely?

(A) 50mL

(B) 100mL

(C) 150mL

(D) 200mL

(E) 400mL

45. Which of the following is not correct for the unbalanced reaction below?

$$Zn(s) + Fe_2O_3(s) \rightarrow ZnO(s) + Fe(s)$$

(A) Oxidation number of Fe in $Fe_2O_3(s)$ is $+3$

(B) Zn is reducing agent

(C) If 3 mole of ZnO(s) is formed, 2 mole of Fe(s) is formed

(D) If 2mole of Zn(s) and 1mole of $Fe_2O_3(s)$ are reacted, $Fe_2O_3(s)$ is limiting reactant

(E) If 3mole of Zn(s) and 1mole of $Fe_2O_3(s)$ are reacted and 1.5mole of Fe(s) is formed, percent yield is 75%

46. When 2mole of $C_3H_8(g)$ reacted with 80.0g of $O_2(g)$ completely in rigid container, which of the following is not correct for the reaction below?

$$C_3H_8(g) + 5O_2(g) \rightarrow 3CO_2(g) + 4H_2O(l)$$

(A) Total pressure of container decreases

(B) $C_3H_8(g)$ is remained after the reaction

(C) The mass of reactant remained is 66.0g

(D) 36.0g of $H_2O(l)$ is produced

(E) The ratio of partial pressure of $C_3H_8(g)$ and $CO_2(g)$ is 1 : 3

47. Which of the following is(are) product(s) of the reaction below?

$$CaCO_3(s) + H_3O^+(aq) \rightarrow$$

I. $CO_2(g)$

II. Ca(s)

III. $H_2O(l)$

(A) I only   (B) II only

(C) I, and II only (D) I, and III only

(E) I, II, and III

48. Which of the following is the example of oxidation and reduction reaction EXCEPT?

(A) Rusting of metal

(B) Decomposition of water to produce hydrogen and oxygen gas

(C) Combustion of propane

(D) Heating of potassium perchlorate to produce the potassium chloride and oxygen gas

(E) Reaction between $AgNO_3(aq)$ and NaCl(aq) to produce the AgCl(s) precipitation

49. Which of the following aqueous solution shows the highest boiling point at same atmospheric pressure condition.

(A) 0.25 molal $Pb(NO_3)_2(aq)$

(B) 0.20 molal $CaCl_2(aq)$

(C) 0.10 molal $C_6H_{12}O_6(aq)$

(D) 0.50 molal $C_{12}H_{22}O_{11}(aq)$

(E) 0.30 molal $NaCl(aq)$

50. Which of the following is correct to increase the amount of product for the reaction in rigid container below?

$$N_2(g) + 3H_2(g) \rightleftharpoons 2NH_3(g) + 92 \text{ kilojoules}$$

I. Add helium gas to container

II. Decrease the temperature

III. Add the $Fe(s)$ catalyst to the container

(A) I only

(B) II only

(C) I, and II only

(D) II, and III only

(E) I, II, and III

51. Which of the following is the largest oxidation state of carbon?

(A) CO

(B) CO_3^{2-}

(C) CH_4

(D) C_2H_2

(E) C_2H_5OH

52. he below is the successive ionization energy of the period 2 elements, A and B

element	Successive Ionization Energy ($E_n, 10^3 kJ/mol$)						
	E_1	E_2	E_3	E_4	E_5	E_6	E_7
(A)	1.4	2.9	4.6	7.5	9.4	53.3	64.4
(B)	1.3	3.4	5.3	7.5	11.0	13.3	71.3

Which of the following is(are) correct for the elements (A), and (B)?

I. The radius of the ion with the most stable electron configuration (B) is greater than (A).

II. Oxide of element (A) is acidic in aqueous solution

II. Element B can form a ionic compound with Na^+ ion in ratio of (B) : Na is 1 : 2.

(A) I only

(B) II only

(C) I, and II only

(D) II, and III only

(E) I, II, and III

53. 125 g of copper sulfate hydrate($CuSO_4 \cdot xH_2O$) is heated to evaporate all of the water, resulting in a mass reduction of 42g. Which of the following is correct value of x?(molar mass of $CuSO_4$ is 180g/mol)

(A) 2

(B) 3

(C) 4

(D) 5

(E) 6

54. Which of the following is the correct pair of conjugate acid and base in the reaction represented below?

$$CO_3^{2-}(aq) + H_2O(l) \rightarrow HCO_3^-(aq) + OH^-(aq)$$

	conjugate acid	conjugate base
(A)	CO_3^{2-}	HCO_3^{-}
(B)	CO_3^{2-}	H_2O
(C)	H_2O	H_3O^{+}
(D)	HCO_3^{-}	CO_3^{2-}
(E)	HCO_3^{-}	H_2O

55. Which of the following reaction shows the greatest decrease in entropy?

(A) $2NH_3(aq) \rightarrow N_2(g) + 3H_2(g)$

(B) $CO_2(g) + H_2O(l) \rightarrow H_2CO_3(aq)$

(C) $[Ag(NH_3)_2]^{+}(aq) \rightarrow Ag^{+}(aq) + 2NH_3(aq)$

(D) $N_2O_4(g) \rightarrow 2NO_2(g)$

(E) $2NO_2(g) \rightarrow 2NO(g) + O_2(g)$

56. Which of the following is not a redox reaction?

(A) $2NaBr(aq) + Cl_2(aq) \rightarrow 2NaCl(aq) + Br_2(aq)$

(B) $CaCO_3(s) \rightarrow CaO(s) + CO_2(g)$

(C) $Zn(s) + 2HCl(aq) \rightarrow ZnCl_2(aq) + H_2(g)$

(D) $2H_2O(aq) \rightarrow 2H_2(g) + O_2(g)$

(E) $Al_2O_3(s) + 3CO(g) \rightarrow 2Al(s) + 3CO_2(g)$

57. Which of the following is the greatest molar concentration of total ions? When the same mass is dissolved in water to make a 1L of aqueous solution.

(A) $NaNO_3$ (molar mass 85g/mol)

(B) NaF (molar mass 42g/mol)

(C) CaF_2 (molar mass 78g/mol)

(D) $CaCl_2$ (molar mass 111g/mol)

(E) $Al(NO_3)_3$ (molar mass 213g/mol)

58. Which of the following is correct for the conversion of nitrogen and nitrogen compounds?

$$N_2$$
$$\downarrow (1)$$
$$NH_3$$
$$\downarrow (2)$$
$$NO_2$$
$$\downarrow (3)$$
$$N_2O_5$$

I. In the conversion (2) and (3), nitrogen is reduced.

II. The molecular geometry of NH_3 is trigonal planar.

III. In the conversion (1), oxidation number of nitrogen decrease from 0 to –3

(A) I only

(B) III only

(C) I, and II only

(D) II, and III only

(E) I, II, and III

59. Which of the following is the correct name for the compound?

(A) Na_2O disodium monoxide

(B) $FeCl_3$ iron chloride

(C) N_2O_4 nitrogen tetroxide

(D) $CuNO_3$ copper(I) nitrate

(E) CO   carbon oxide

60. Which of the following is correct set for the molecular geometry, polarity and hybrid orbital of central atom for the given compound?

(A) CO_2 linear   polar sp^2

(B) NH_3 trigonal pyramidal   nonpolar sp^3

(C) CH_4 square planar   polar sp^3

(D) CCl_4 tetrahedral   nonpolar sp^3

(E) H_2O bent(V-shape)   polar sp^2

61. The below is unbalanced combustion equation of butane.

$$...C_4H_{10}(g) + ...O_2(g) \rightarrow ...CO_2(g) + H_2O(l)$$

which of the following is correct for the reaction represented above?
(the relative atomic mass of H : 1g/mol, C : 12g/mol, O : 16g/mol)

I. The sum of the reaction coefficients of reactants is larger than that of the products

II. When 1 mol of $C_4H_{10}(g)$ and 10 mol of $O_2(g)$ are reacted to produce 4 mol of water, the % yield is 80%

III. When 29g of $C_4H_{10}(g)$ is reacted with excess $O_2(g)$ completely, the volume of $CO_2(g)$ produced is 44.8L at STP(0℃, 1atm)

(A) I only   (B) II only
(C) I and II only   (D) II and III only
(E) I, II, and III

62. The belows are the thermochemical equations associated with the combustion reaction of carbon.

$$2C(s, graphite) + O_2(g) \rightarrow 2CO(g)$$
$$\Delta H = -566kJ$$
$$C(s, graphite) + O_2(g) \rightarrow CO_2(g)$$
$$\Delta H = -394kJ$$

Which of the following is correct for the enthalpy change, ΔH, for the reaction represented below?

$$CO(g) + 1/2O_2(g) \rightarrow CO_2(g)$$

(A) −111kJ   (B) +111kJ
(C) −222kJ   (D) +222kJ
(E) −333kJ

63. Which of the following conditions is the highest density of Ar(g)?(molar mass : 40g/mol)

(A) 273K, 1atm
(B) 373K, 2atm
(C) 373K, 1atm
(C) 473K, 2atm
(D) 546K, 2atm

64. The half-life of the radioactive isotope ^{226}Ra is 1,600 years. The below is the decay reaction of ^{226}Ra.

$$^{226}_{88}Ra \rightarrow ^{222}_{86}Rn + \boxed{}$$

Which of the following is correct for the blank and mole ratio of Ra : Rn after 4,800 years?

(A) $^{4}_{2}He$ 1 : 3
(B) $^{-1}_{0}e$ 1 : 3
(C) $^{4}_{2}He$ 1 : 7
(D) $^{0}_{0}\gamma$ 1 : 7
(E) $^{+1}_{0}e$ 1 : 7

65. To determine the density of the metal, volume and mass are measured by placing it in a graduated cylinder containing 100.0cm³ of water. the data collected from experiment is represented below.

The mass before adding the metal : 600.42g
The mass of after adding the metal : 652.00g
The volume after adding the metal : 140.0cm³

which of the following is correct for the density of metal to be reported?

(A) $1.290 g/cm^3$
(B) $1.30 g/cm^3$
(C) $1.2895 g/cm^3$
(D) $1.3 g/cm^3$
(E) $1 g/cm^3$

II. The partial pressure of NO(g) is 4atm after the reaction

(A) I only
(B) II only
(C) I and II only
(D) II and III only
(E) I, II, and III

66. Which of the following is amphoteric species?

(A) NH_3
(B) H_2CO_3
(C) HPO_4^{2-}
(D) CO_3^{2-}
(E) NO_3^{-}

67. Which of the following is basic in aqueous solution?

(A) NH_4Cl
(B) Na_3PO_4
(C) $Ca(NO_3)_2$
(D) Na_2SO_4
(E) $CaBr_2$

69. Which of the following is correct for the experiment safety EXCEPT?

(A) All chemicals must be handled carefully and in a fume hood whenever possible

(B) Long-sleeved shirts and leather-topped shoes should not be worn at all times

(C) Flush with large quantities of water when disposing of liquid chemicals or solutions in the sink

(D) Glass tubing should be forced into a rubber stopper to seal the tube perfectly.

(E) All glassware must be washed and cleaned

68. $N_2(g)$ and $O_2(g)$ are contained in the same volume of flask as shown below.

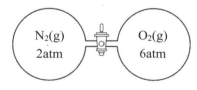

$N_2(g)$ and $O_2(g)$ react completely to form NO gas after opening the coke. Which of the following is correct for the experiment?(the volume of the connector is ignored and temperature remains the same before and after the reaction)

I. The mole ratio of $N_2(g) : O_2(g)$ is 1 : 3 before the reaction

II. $O_2(g)$ and $N_2(g)$ remain after the reaction

70. The table represented below is the temperature, pressure and volume of $H_2(g)$ and $O_2(g)$

Gas	Temperature	pressure	volume
$H_2(g)$	200K	4atm	50L
$O_2(g)$	800K	8atm	100L

Which of the following is correct?

(A) The number of moles of hydrogen and oxygen gas is the same

(B) The average kinetic energy of $O_2(g)$ is the same as that of $H_2(g)$

(C) The density of $O_2(g)$ is sixteen times that of $H_2(g)$

(D) The average speed of $H_2(g)$ is slower than that of $O_2(g)$

(E) All the $H_2(g)$ molecules are faster than $O_2(g)$ molecules

Note : For all questions involving solution, assume that the solvent is water unless otherwise stated. Through the test the following symbols have the definitions specified unless otherwise noted.

H	=	enthalpy	atm	=	atmosphere(s)
M	=	molar	g	=	gram(s)
n	=	number of moles	J	=	joule(s)
P	=	pressure	kJ	=	kilojoule(s)
R	=	molar gas constant	L	=	liter(s)
S	=	entropy	mL	=	milliliter(s)
T	=	temperature	mm	=	millimeter(s)
V	=	volume	mol	=	mole(s)
V	=	volt(s)			

Part A

Directions : Each set of lettered choices below refers to the numbered statements or questions immediately following it. Select the one lettered choice that best fits each statement of answers each question and then fill in the corresponding circle on the answer sheet. A choice may be used once, more than once, or not at all in each set.

Questions 1-4 refer to the following elements

(A) Sodium   (B) Iron
(C) Chlorine   (D) Carbon
(E) Helium

1. is a diatomic molecule at thermodynamic standard state, 25℃, 1atm

2. exists in various oxidation states in ionic compounds

3. reacts with water violently to produce hydrogen gas.

4. does not need to form chemical bonds to have a stable electron configuration. arrangement.

Questions 5-8 refer to the following reactions

(A) Neutralization reaction
(B) Single replacement reaction
(C) Combustion reaction
(D) Precipitation reaction
(E) Decomposition reaction

5. $H_2SO_4(aq) + Na_2O(s) \rightarrow H_2O(l) + Na_2SO_4(aq)$

6. $CaCO_3(s) \rightarrow CaO(s) + CO_2(g)$

7. $Zn(s) + CuSO_4(aq) \rightarrow ZnSO_4(aq) + Cu(s)$

8. $Pb(NO_3)_2(aq) + Na_2S(aq) \rightarrow 2NaNO_3(aq) + PbS(s)$

Questions 9-13 refer to the following substances

(A) H_2O

(B) CO_2

(C) NH_3

(D) O_3

(E) $CHCl_3$

9. is a nonpolar molecule.

10. is a linear molecule.

11. is a polar molecule with resonance structures.

12. is a polar molecule with tetrahedron structure.

13. is very soluble in water but only slightly ionized.

Questions 14-17 refer to the following elements

(A) Oxygen

(B) Fluorine

(C) Sodium

(D) Magnesium

(E) Manganese

14. is diamagnetic

15. has an electron configuration of the partially filled d orbitals at the ground states.

16. has the highest first ionization energy.

17. has the smallest atomic radius.

Questions 18-20 refer to the following definition

(A) London dispersion force

(B) Dipole-Dipole attraction

(C) Hydrogen bond

(D) Ionic bond

(E) Metallic bond

18. is the intermolecular force present in all molecules

19. is a type of chemical bonding that arises from the electrostatic attractive force between delocalized free electron and positively charged ion

20. does not conduct current in the solid state, but it can flow current if it dissolves in water

Questions 21-25 refer to the following 1M of aqueous solutions

(A) $NaCl(aq)$ (B) $NaHCO_3(aq)$

(C) $Ca(NO_3)_2$ (D) $NH_4ClO_4(aq)$

(E) $CuSO_4(aq)$

21. is weakly basic

22. reacts with the $AgNO_3(aq)$ to form a white precipitate

23. can react with an acid or a base

24. is the most acidic

25. reacts with $Zn(s)$ to change the color of the solution.

PLEASE GO TO THE SPECIAL SECTION AT THE LOWER LEFT-HAND CORNER OF THE PAGE OF THE ANSWER SHEET YOU ARE WORKING ON AND ANSWER QUESTIONS 101-115 ACCORDING TO THE FOLLOWING DIRECTION.

Part B

Direction: Each question below consists of two statements. I in the left-hand column and II in the right-hand column. For each question, determine whether statement I is true of false <u>and</u> whether statement II is true or false and fill in the corresponding T or F circles on your answer sheet. <u>Fill in circle CE only if statement II is a correct explanation of the true statement I</u>

EXAMPLE:

	I		II
EX1.	HNO$_3$ is a strong acid	Because	HNO$_3$ contains nitrogen.
EX2.	An atom of fluorine is electrically neutral.	Because	an fluorine contains an equal number of protons and electrons.

SAMPLE ANSWER

	I	II	CE*
EX1	● Ⓕ	● Ⓕ	○
EX2	● Ⓕ	● Ⓕ	●

	I		II
101.	As the temperature increases, the reaction rate increases	Because	The ratio of the number of particles having an activation energy or more increases at higher temperature
102.	The average speed of SO$_2$ and He is the same at the same temperature	Because	The average kinetic energy of ideal gas depends only on temperature
103.	Allotropes have the same chemical properties	Because	The allotrope consists of the same element, but the manner of bonding is different
104.	The pH of 1.0×10^{-5}M of NaOH(aq) is 9 at 25℃	Because	NaOH is strong base and autoionization constant of water is 1.0×10^{-14} at 25℃
105.	The first ionization energy of F is larger than that of Cl	Because	F is more electronegative than Cl.

106. The volume of the molar gas may be greater than 22.4 L | Because | The ideal gas that has mass but no particle itself volume is not a real substance

107. $NH_4Cl(s)$ is soluble in water | Because | There are both covalent and ionic bonds in $NH_4Cl(s)$

108. Nitrogen and oxygen have similar chemical properties. | Because | Elements in the same period have the same valence shell in ground state.

109. $HCl(aq)$ is the acid of Brønsted -Lowry but not of Arrhenius | Because | The definition of Arrhenius includes the definition of Brønsted-Lowry

110. In the equilibrium reaction $N_2(g) + O_2(g) \rightleftharpoons 2NO(g)$, reducing the volume of the reaction vessel can increase the yield of products from the same amount of reactants. | Because | If the volume of the vessel is reduced in the equilibrium reaction, the equilibrium shifts in the direction of decreasing the sum of the gas mole numbers.

111. CH_4 is a nonpolar molecule. | Because | The molecules are symmetrical and the dipoles of the bond are canceled.

112. The boiling point of CH_3CH_2OH, ethanol, is higher than that of dimethyl ether, CH_3OCH_3 | Because | Alcohols and ethers of the same carbon number are structural isomers.

113. Chromatography is a method of separating the mixture | Because | Chromatography is a method using the difference in affinity between the mobile phase and the stationary phase of the mixed component.

114. Entropy increases in the endothermic reaction. | Because | The sum of bond (dissociation) energies in the product is greater than in the reactant in the endothermic reaction

115. Ammonia can be obtained from water displacement | Because | Ammonia forms hydrogen bonds with water molecules.

Part C

Direction: Each of the questions or incomplete statements below is followed by five suggested answer or completions. Select the one that is best in each case and then fill in the corresponding circle on the answer sheet.

26. Which of the following is not correct Lewis structure?

(A)

(B)

(C) $H—C\equiv C—H$

(D)

(E)

27. $N_2(g) + O_2(g) \rightarrow 2NO(g)$ $\Delta H = 180kJ$

$N_2(g) + 3H_2(g) \rightarrow 2NH_3(g)$ $\Delta H = -92kJ$

$2H_2(g) + O_2(g) \rightarrow 2H_2O(g)$ $\Delta H = -484kJ$

Which of the following is correct for the enthalpy change in kJ of the reaction represented below?

$$4NH_3(g) + 5O_2(g) \rightarrow 4NO(g) + 6H_2O(g)$$

(A) -1996 (B) -1172

(C) $+1172$ (D) -908

(E) $+908$

28. The below is the unbalanced reaction equation

$$...SiO_2 + ...HF \rightarrow ...H_2O + ...H_2SiF_6$$

Which of the following is(are) the correct for the reaction above?

I. The sum of the reaction coefficients as the whole number is 10

II. It is neutralization reaction

III. Acid can not be stored in glass bottle

(A) I only   (B) II only

(C) I, and III only   (D) II, and III only

(E) I, II, and III

29. The below is the unbalanced reaction equation

$$...H_2(g) + ...NO(g) \rightarrow ...H_2O(g) + ...N_2(g)$$

Which of the following is correct for the reaction represented above when 1mol of $H_2(g)$ are reacted with 2mol of $NO(g)$ in rigid container at constant temperature?

(A) The sum of the reaction coefficients as the whole number is 8

(B) $H_2(g)$ is oxidizing agent

(C) Forward reaction is the reaction of increasing entropy.

(D) If the limiting reactant is consumed completely, the total pressure of the container increases

(E) If the % Yield is 70%, the number of moles present in the container after the reaction is 2.65 mol

30. Which of the following is the largest number of unpaired electrons in the p orbitals?

(A) Carbon

(B) Nitrogen

(C) Oxygen

(D) Fluorine

(E) Argon

31. Which of the following is not correct for the electron configuration of the period 4 elements at the ground state?

(A) $1s^2 2s^2 2p^4 3s^2 3p^6 4s^2$

(B) $1s^2 2s^2 2p^4 3s^2 3p^6 3d^4 4s^2$

(C) $1s^2 2s^2 2p^4 3s^2 3p^6 3d^6 4s^2$

(D) $1s^2 2s^2 2p^4 3s^2 3p^6 3d^{10} 4s^1$

(E) $1s^2 2s^2 2p^4 3s^2 3p^6 3d^{10} 4s^2 4p^2$

32. Which of the following is correct explanation for the kinetics?

(A) Increasing the temperature increases the reaction rate of forward reaction but decrease the reaction rate of reverse reaction in endothermic reaction

(B) Increasing the concentration of the reactant increases the yield of the product

(C) Catalyst can alter the reaction mechanism and increase the reaction rate by changing the heat of reaction

(D) Even when the catalyst is used, the energy distribution of the reactants does not change

(E) The rate of a reaction with a large enthalpy change is slow

33. Which of the following is the smallest number of oxygen atoms?

(A) 0.20mol of H_2CO_3

(B) 6.0×10^{22} particles of H_2SO_4

(C) 0.15mol of H_3PO_4

(D) 8.96L of CO at 0℃, 0.5atm

(E) 6.72L of NO at 273℃, 4.0atm

34. When 20.0mL of 0.1M $HC_2H_3O_2$(aq) is titrated with 0.05M NaOH(aq), which of the following is not correct for the reaction at 25℃?

(A) The volume of NaOH added to the equivalent point is 40.0mL.

(B) pH at equilibrium point is higher than 7.00

(C) Phenolphthalein is suitable as an indicator.

(D) The concentration of Na^+ and $C_2H_3O_2^-$ at the equivalent point is the same.

(E) The mole number of Na^+ is 0.002mol at the equivalent point

35. A 5L, 2atm of N_2 gas, 4L, 4atm of O_2 gas and 8L, 3atm of He gas are mixed in a 10L vessel. The gases do not react with each other. Which of the following is correct for the partial pressure of each gas in the mixture?

	P_{N2}	P_{O2}	P_{He}
(A)	1.0atm	2.4atm	1.6atm
(B)	2.4atm	1.6atm	1.0atm
(C)	1.0atm	1.6atm	2.4atm
(D)	2.4atm	1.0atm	1.6atm
(E)	1.6atm	2.4atm	1.0atm

36. The below is the unbalanced gaseous reaction

$$...CH_4(g) + ...H_2O(g) \rightarrow ...CO(g) + ...H_2(g)$$

What is the volume of $H_2(g)$ produced of when 0.5mol of $CH_4(g)$ and 1mol of $H_2O(g)$ react completely at $0℃$, 1atm?

(A) 5.60L   (B) 11.2L

(C) 22.4L   (D) 33.6L

(E) 44.8L

[37~38] The below is the decomposition reaction of the ester under basic conditions and initial rate of reaction is represented in table

$$RCOOR'(aq) + OH^-(aq)$$
$$\rightarrow RCOO^-(aq) + ROH(aq)$$

Ex	Initial concentration of ester(M)	Initial concentration of hydroxide(M)	Initial rate (M/s)
1	0.020	0.030	$6.0×10^{-4}$
2	0.020	0.045	$9.0×10^{-4}$
3	0.030	0.060	$1.8×10^{-3}$

rate law : $k[RCOOR]^m[OH^-]^n$

k : rate constant, m, n : reaction order

37. Which of the following is correct for the reaction oder with respect to [RCOOR'], [OH⁻] and overall reaction?

	[RCOOR']	[OH⁻]	overall
(A)	zero	first	first
(B)	zero	second	second
(C)	first	first	second
(D)	first	second	third
(E)	second	first	third

38. Which of the following is correct for the rate constant(k)?

(A) $1.0×10^{-1}s^{-1}$

(B) $1.0×10^{-1}M^{-1}s^{-1}$

(C) $1.0M^{-1}s^{-1}$

(D) $1.0×10^{-1}M^{-2}s^{-1}$

(E) $1.0M^{-2}s^{-1}$

39. 10 mL of concentrated $H_2SO_4(aq)$ is diluted with distilled water to make a 100 mL aqueous solution. 400mL of 0.1M NaOH(aq) is consumed to titrate diluted $H_2SO_4(aq)$ completely Which of the following is not correct for the experiment at $25℃$?

(A) The concentration of concentrated $H_2SO_4(aq)$ is 2M

(B) The pH of mixture at equivalent point is 7.0

(C) indicator is added to titrant before the titration

(D) Phenolphthalein is appropriate indicator

(E) The concentration of $Na^+(aq)$ is 0.08M at equivalent point

40. Which of the following is correct for the changing of electrical conductivity of the mixture with the volume of added NaOH(aq) when 200mL of 0.2M HCl is titrated with 0.1M NaOH(aq)?

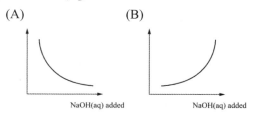

(A)   (B)

NaOH(aq) added   NaOH(aq) added

(C)   (D)

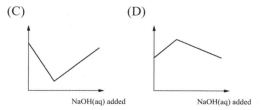

(E)

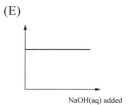

41. Which of the following gases shows largest density?

(A) $0℃$, 0.5atm $SO_2(g)$

(B) $0℃$, 1atm $N_2(g)$

(C) $273℃$, 1atm $O_3(g)$

(D) $273℃$, 2atm $Ar(g)$

(E) $273℃$, 2atm $CO_2(g)$

42. Which of the following is not for the conjugate acid and base pair?

(A) NH_4^+ and NH_3

(B) H_2CO_3 and CO_3^{2-}

(C) H_2O and OH^-

(D) $H_2PO_4^-$ and HPO_4^{2-}

(E) H_3O^+ and H_2O

43. The below is the unbalanced decomposition reaction of $Cu_2O(s)$

$$...Cu_2O(s) \rightarrow ...Cu(s) + ...O_2(g)$$

Which of the following is not correct explanation for the reaction?(molar mass of Cu : 63.5g/mol, O : 16g/mol)

(A) Oxidation number of Cu in Cu_2O is $+1$

(B) Entropy increases when forward reaction occurs

(C) The number of electrons transferred per 1mol of Cu_2O is 2mol

(D) The mass of $O_2(g)$ produced when 1mol of $Cu(s)$ is produced is $(32/63.5) \times 2$g

(E) The volume of O_2 (g) produced when 1 mol of Cu_2O is completely decomposed at $25℃$, 1atm is larger than 11.2L

44. $...[Cu(NH_3)_2]^+ + ...NH_3(aq) + ...H_2O(l) + ...O_2(g) \rightarrow [Cu(NH_3)_4]^{2+} + OH^-(aq)$

Which of the following is the correct for sum of the reaction coefficients when the equation is balanced as the simplest whole number?

(A) 20

(B) 21

(C) 22

(D) 23

(E) 24

45. Which of the following is correct for the ionization energy?

(A) $I_2(s) \rightarrow 2I^+(s) + 2e^-$

(B) $I_2(s) \rightarrow 2I^+(g) + 2e^-$

(C) $I(g) \rightarrow I^+(g) + e^-$

(D) $I_2(g) + e^- \rightarrow 2I(g)^-$

(E) $I_2(s) + e^- \rightarrow 2I(g)^-$

46. The below is the HI(g) formation reaction from $H_2(g)$ and $I_2(g)$

$$H_2(g) + I_2(g) \rightleftharpoons 2HI(g) \qquad \Delta H > 0$$

The equilibrium constant(K_C) for the formation of HI(g) is 40 at certain temperature. Which of the following is correct explanation for the reaction?

(A) The sum of potential energy of reactant is larger than that of product

(B) When $[H_2] = 0.1M$, $[I_2] = 0.001M$ $[HI] = 0.1M$ the reaction shift to reverse reaction

(C) When a noble gas is added reactants are more favorable than products

(D) When the temperature increases reactants are more favorable

(E) When HI is added in the equilibrium state, the equilibrium shift to reverse reaction and the equilibrium constant decreases.

47. Which of the following is correct for number of sigma(σ) and pi(π) bonds in below?

$$HOOC-(CH_2)_4-CONH-(CH_2)_6-NH_2$$

	sigma(σ) bond	pi(π) bond
(A)	38	4
(B)	39	3
(C)	40	2
(D)	41	1
(E)	42	0

48. Which of the following is correct for the number of d orbital electrons in Fe of $Fe(NO_3)_3$?

(A) 3   (B) 4   (C) 5

(D) 6   (E) 7

49. The mole fractions of He(g) and Ar(g) in the 27℃ 5L vessel are 0.4 and 0.6, respectively. What is the correct expression of the mole number of He(g) in the vessel when the total pressure of gas is 2atm?

(A) $\dfrac{22.4 \times 4}{273 \times 300}$ mol

(B) $\dfrac{273 \times 4}{22.4 \times 300}$ mol

(C) $\dfrac{273 \times 2}{22.4 \times 300}$ mol

(D) $\dfrac{300 \times 6}{22.4 \times 273}$ mol

(E) $\dfrac{273 \times 0.4}{22.4 \times 300}$ mol

50. The belows are the standard reduction potentials of $Ag^+(aq)$ and $Cu^{2+}(aq)$. What is the correct standard cell potential in Galvanic cell using the Ag(s) and Cu(s)?

$Ag^+(aq) + e^- \rightarrow Ag(s)$	$E^o_{red} = +0.80V$
$Cu^{2+}(aq) + 2e^- \rightarrow Cu(s)$	$E^o_{red} = +0.34V$

(A) $+1.94V$ (B) $+1.26V$

(C) $+0.46V$ (D) $-1.26V$

(E) $-0.46V$

51. Which of the following is correct?

I. Boiling point at 1atm
0.1m of $C_{12}H_{22}O_{11}(aq)$ < 0.07m of NaCl(aq)

II. Vapor pressure at same temperature
0.05m of $Al_2(SO_4)_3$ < 0.1m of $Na_2SO_4(aq)$

III. Osmotic pressure
20g/L of $C_6H_{12}O_6(aq)$ < 20g/L of $C_5H_{10}O_5(aq)$

(A) I only   (B) II only

(C) I, and II only   (D) I and III only

(E) I, II, and III

52. The below is the unbalanced reaction between Zn(s) and HCl(aq)

$$...Zn(s) + ...HCl(aq) \rightarrow ...ZnCl_2(aq) + ...H_2(g)$$

Which of the following is correct for the reaction? when Zn(s) is reacted with excess of HCl(aq) completely. and the gas collected in 500mL of bottle over the water (The water line of inside and outside of collecting bottle is the same. The temperature is remained 25℃. The vapor pressure of water is 24mmHg and atmospheric pressure is 754mmHg)

I.   This reaction is oxidation-reduction reaction

II. $H_2(g)$ can be collected on water because of its low solubility in water

III. The dry volume of $H_2(g)$ produced is

$$\frac{730}{754} \times 0.5L$$

(A) I only   (B) II only

(C) I, and II only   (D) II, and III only

(E) I, II, and III

53. Which of the following is correct for the unbalanced reaction represented below?(the reaction coefficients is the simplest whole umber)

$$...NH_3(g) + ...O_2(g) \rightarrow ...NO(g) + ...H_2O(g)$$
$$\Delta H = -918kJ$$

I.   Entropy increases in the forward reaction

II. The oxidation number of nitrogen increases from -3 to $+2$

III. When 12×10^{23} NH_3 molecules react completely 459 kJ of heat is absorbed

(A) I only   (B) II only

(C) I, and II only   (D) II, and III only

(E) I, II, and III

54. Which of the following is not correct for the reaction represented below?

$$2NO_2(aq) + 7H_2(aq) \rightarrow 2NH_3(aq) + 4H_2O(l)$$

(A) The reaction oxidation-reduction reaction

(B) NO_2 is oxidizing agent

(C) 14 mole of electrons are transferred per 1mole of NO_2

(D) When the reaction occurs, the pH increases.

(E) The hybrid orbital of NH_3 and H_2O is the same

55. $N_2(g) + O_2(g) \rightleftharpoons 2NO(g)$ $\Delta H > 0$

Initially 2.00 mol of N_2 and 4.00 mol of $O_2(g)$ are contained in a 0.50 L of evacuated rigid vessel. The concentration of NO(g) at the equilibrium state is 4.00M. Which of the following is not correct for the reaction represented above?

(A) Standard enthalpy of formation of NO(g) is positive

(B) When the reaction occurs, the temperature of surrounding decreases

(C) The equilibrium constant, Kc is 0.5

(D) The sum of bond enthalpy of the products is smaller than that of reactants

(E) When the temperature increases, the equilibrium shifts to the forward reaction and the equilibrium constant increases.

56. The unknown element X reacts with nitrogen(N) to form compounds A and B. The mass ratio of nitrogen to X (N : X) is 7 : 8 for compound A and 7 : 20 for compound B. Which of the following is correct?(in compound A, mol ratio of N : X is 1:1 and molecular formula is the same as empirical formula)

I. The unknown element X is oxygen(O)
II. A and B are acidic when dissolved in water
III. The oxidation number of nitrogen(N) is +4 in compound B

(A) I only
(B) II only
(C) I, and II only
(D) II, and III only
(E) I, II, and III

57. Which of the following is correct reaction condition to increasing the yield of products for the reaction represented below?

$$2NO(g) + Cl_2(g) \rightleftharpoons 2NOCl(g) \qquad \Delta H < 0$$

(A) Low temperature and high pressure
(B) Low temperature and low pressure
(C) High temperature and high pressure
(D) High temperature and low pressure
(E) Regardless of temperature and pressure, the forward reaction is always spontaneous

58. 100℃, 100g of metal was placed in 50.0g of water at 20.0℃. When equilibrium is reached, the final temperature is 35℃. Which of the following is correct for the specific heat capacity(c) ratio between metal and water, c(metal) : c(water)?

(A) 1 : 5
(B) 3 : 13
(C) 3 : 26
(D) 4 : 12
(E) 5 : 11

59. Which of the following is not correct for the intermolecular force?

(A) London dispersion force(LDF) exists in all molecules
(B) The normal boiling point of a polar molecule is not always greater than that of a nonpolar molecule
(C) The boiling points of the structural isomers are different.
(D) HCl(hydrogen chloride) has a higher normal boiling point than HBr(hydrogen bromide) because the dipole-dipole force of HCl is larger than that of HBr
(E) The normal boiling point of acetic acid is measured higher than expected because acetic acid forms a dimer through hydrogen bonding

60. The below is the autoionization reaction of water

$$H_2O(l) + H_2O(l) \rightleftharpoons H_3O^+(aq) + OH^-(aq)$$
$$\Delta H > 0, \ Kw = 1.0 \times 10^{-14}$$

Which of the following is correct?

(A) The temperature increases, the pH of the water increases and becomes basic
(B) The temperature increases, the pH of the water decreases and becomes basic
(C) The temperature decreases, the pH of the water increases and becomes basic
(D) The temperature decreases, the pH of the water decreases and becomes basic

(E) The pH of water can vary with temperature but is always neutral

61. The diffusion rate of XO_2, an oxide of the unknown element X at same temperature, is half of that of CH_4. Which of the following is correct?

I. The average kinetic energy of CH_4 and XO_2 is the same

II. Average speed of molecule is proportional to the Kelvin temperature

III. Unknown element, X can be S(sulfur)

(A) I only

(B) II only

(C) I, and II only

(D) I, and III only

(E) I, II, and III

62. $CH_4(g)$ and $O_2(g)$ are placed in a 2.0L of evacuated rigid container and then completely burned. Total pressure is 5.0 atm and mole fraction of $CH_4(g)$ is 0.2 before the reaction. Which of the following is not correct for the reaction is completed?(the temperature remains constant)

$$...CH_4(g) + ...O_2(g) \rightarrow ...CO_2(g) + ...H_2O(l)$$

(A) Partial pressure of $O_2(g)$ is 4.0 atm before the reaction

(B) Forward reaction is a reaction in which the entropy decreases

(C) Limiting reactant is $CH_4(g)$

(D) Total pressure is 3.0 atm after the reaction

(E) Mole ration of $CO_2(g)$ in gaseous components is 1/2

63. Which of the following is not correct?

(A) HCl(aq) is strong electrolyte

(B) The pH of $10^{-8}M$ of $HNO_3(aq)$ is less than 7 at 25℃

(C) $H_2SO_4(aq)$ can reacts with Zn(s) to produce $H_2(g)$ but $CH_3COOH(aq)$ can not

(D) 2.0L of 0.5M HCl(aq) is needed to neutralize 1.0L of 1.0M $NH_3(aq)$ and the pH at equivalent point is less than 7 at 25℃

(E) The volume of 0.1M NaOH (aq) required to neutralize the same concentration and volume of hydrochloric acid and acetic acid respectively is the same.

64. The belows are the standard reduction potentials for some species

$Fe^{3+}(aq) + e^- \rightarrow Fe^{2+}(aq)$	$E^o_{red} = +0.77V$
$Cu^+(aq) + e^- \rightarrow Cu(s)$	$E^o_{red} = +0.52V$
$I_2(s) + 2e^- \rightarrow 2I^-(aq)$	$E^o_{red} = +0.54V$

Which of the following is strongest reducing agent?

(A) Fe^{3+} (B) Fe^{2+}

(C) Cu   (D) I_2

(E) $2I^-$

65. Generally, as the temperature increases, the reaction rate increases. This is because

(A) activation energy of forward reaction decreases

(B) transition state becomes stable

(C) the average energy of the reactants increases and mole fraction of particles with or more than activation energy increase

(D) heat of reaction decrease

(E) the reaction proceeds alternative pathway and mole fraction of molecules with or more than activation energy increase

66. Which of the following compounds has the smallest oxidation number of hydrogen?

(A) H_2O_2

(B) $HClO_4$

(C) HCN

(D) H_2O

(E) KH

67. The below is the unbalanced reaction between Na(s) and $H_2O(l)$

$$Na(s) + H_2O(l) \rightarrow NaOH(aq) + H_2(g) \quad \Delta H < 0$$

Which of the following is not correct for the reaction?

(A) The forward reaction is always spontaneous

(B) The reaction is oxidation and reduction reaction

(C) When 11.5g of Na(s) react with water completely, 5.60L of $H_2(g)$ are produced at STP(0℃, 1atm)

(D) 2 mol of electrons are transferred to $H_2O(l)$ when 1mol of Na(s) are oxidized

(E) The pH of the solution increases after the reaction

68. The below is the reaction for the production of syngas from CO(g) and $H_2(g)$

$$CO(g) + 3H_2(g) \rightleftharpoons CH_4(g) + H_2O(g) \quad \Delta H < 0$$

Which of the following is correct ?

(A) The forward reaction is spontaneous at high temperature

(B) The temperature should be lowered as much as possible in order to increase the yield, but the reaction may not occur at too low temperature

(C) When the temperature increases, the forward reaction rate decreases but the reverse reaction rate increases

(D) The use of catalysts can increase the amount of product

(E) When Ar(g) is added at the constant pressure, the equilibrium shifts to the forward reaction

69. Which of the following compounds can be formed from Al(s) and $N_2(g)$?

(A) AlN

(B) Al_2N_3

(C) Al_3N_4

(D) Al_5N_2

(E) Al_2N_5

70. Which of the following is correct for the density of Na(s) when the volume of 9.5231g of Na(s) is 12.4cm^3?

(A) 0.8 g/cm^3

(B) 0.77 g/cm^3

(C) 0.768 g/cm^3

(D) 0.7679 g/cm^3

(E) 0.76799 g/cm^3

Chemistry Test No.5

Note : For all questions involving solution, assume that the solvent is water unless otherwise stated. Through the test the following symbols have the definitions specified unless otherwise noted.

H	=	enthalpy		atm	=	atmosphere(s)
M	=	molar		g	=	gram(s)
n	=	number of moles		J	=	joule(s)
P	=	pressure		kJ	=	kilojoule(s)
R	=	molar gas constant		L	=	liter(s)
S	=	entropy		mL	=	milliliter(s)
T	=	temperature		mm	=	millimeter(s)
V	=	volume		mol	=	mole(s)
V	=	volt(s)				

Part A

Directions : Each set of lettered choices below refers to the numbered statements or questions immediately following it. Select the one lettered choice that best fits each statement of answers each question and then fill in the corresponding circle on the answer sheet. A choice may be used once, more than once, or not at all in each set.

Questions 1-4 refer to the following substance at 25℃, 1atm

(A) H_2O

(B) CO_2

(C) Au

(D) NH_4Cl

(E) He

1. has both covalent and ionic bond

2. is a molecule that dissolves in water shows acidity

3. is monoatomic molecule

4. exhibits malleability and ductility

Questions 5-8 refer to the following reactions

(A) Arrhenius acid reaction

(B) Lewis acid and base reaction

(C) Precipitation reaction

(D) Oxidation and reduction reaction

(E) Decomposition reaction

5. $AgNO_3(aq) + NaCl(aq) \rightarrow NaNO_3(aq) + AgCl(s)$

6. $NaBr(aq) + Cl_2(aq) \rightarrow NaCl(aq) + Br_2(aq)$

7. $HCN(aq) \rightleftharpoons H^+(aq) + CN^-(aq)$

8. $NH_3(g) + BF_3(g) \rightarrow NH_3BF_3(s)$

Questions 9-11 refer to the following compounds

(A) HCN
(B) CO_2
(C) NH_3
(D) CH_4
(E) $BeCl_2$

9. is a linear polar molecule

10. is nonpolar molecule with two pi(π) bonds

11. has the highest oxidation number of the central atom

Questions 12-15 refer to the following compounds

(A) NaCl
(B) Na_2O
(C) Ca_3N_2
(D) CO_2
(E) Al_2O_3

12. can react with both acid and base

13. can react with $Ca(OH)_2$(aq) and form a white precipitate

14. contains the element with the largest positive oxidation number

15. is the salt from the neutralization between strong acid and strong base

Questions 16-18 refer to the following elements

(A) Noble gases
(B) Halogens
(C) Oxygen group elements
(D) Alkali metals
(E) Transition elements

16. contains the elements with three phases ; solid, liquid and gas at thermodynamic standard state($25°C$, 1atm)

17. can form a complex ion with various oxidation state

18. are all shiny, soft, highly reactive metals at standard temperature and pressure and readily lose their outermost electron to form cations with charge $+1$

Questions 19-21 refer to the following reaction represented by the following equation

(A) $AgNO_3$(aq) $+ 2NH_3$(aq)$\rightarrow[Ag(NH_3)_2]^+(aq)+ NO_3^-$(aq)
(B) C_3H_8(g)$+ 5O_2$(g)$\rightarrow 3CO_2$(g)$+ 4H_2O$(l)
(C) O_2(g)$+ 4H^+$(aq)$+ 4e^-\rightarrow 2H_2O$(l)
(D) Pb^{2+}(aq)$+ S^{2-}$(aq)$\rightarrow PbS$(s)
(E) Na(g)$\rightarrow Na^+$(aq)$+ e^-$

19. A combustion reaction

20. Precipitation reaction

21. A half$-$reaction of reduction

Questions 22-23 refer to the following pieces of laboratory equipment

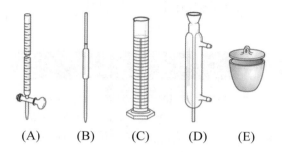

(A)   (B)   (C)   (D)   (E)

22. can be used to collect vapors by condensing them into liquid as they contact the liquid-cooled inner surface of the condenser.

23. is a container in which metals or other substances can be heated to very high temperatures.

Questions 24-25 refer to energy changes associated with the following processes

(A) $2K(s) + 1/2O_2(g) \rightarrow K_2O(s)$

(B) $CH_4(g) + O_2(g) \rightarrow CO_2(g) + H_2O(l)$

(C) $H^+(aq) + OH^-(aq) \rightarrow H_2O(l)$

(D) $2NO_2(g) \rightarrow 2NO(g) + O_2(g)$

(E) $NaCl(s) \rightarrow Na^+(aq) + Cl^-(aq)$

24. Heat of neutralization between strong acid and strong base

25. Heat of formation

PLEASE GO TO THE SPECIAL SECTION AT THE LOWER LEFT-HAND CORNER OF THE PAGE OF THE ANSWER SHEET YOU ARE WORKING ON AND ANSWER QUESTIONS 101-115 ACCORDING TO THE FOLLOWING DIRECTION.

Part B

Direction: Each question below consists of two statements. I in the left-hand column and II in the right-hand column. For each question, determine whether statement I is true of false <u>and</u> whether statement II is true or false and fill in the corresponding T or F circles on your answer sheet. <u>Fill in circle CE only if statement II is a correct explanation of the true statement I</u>

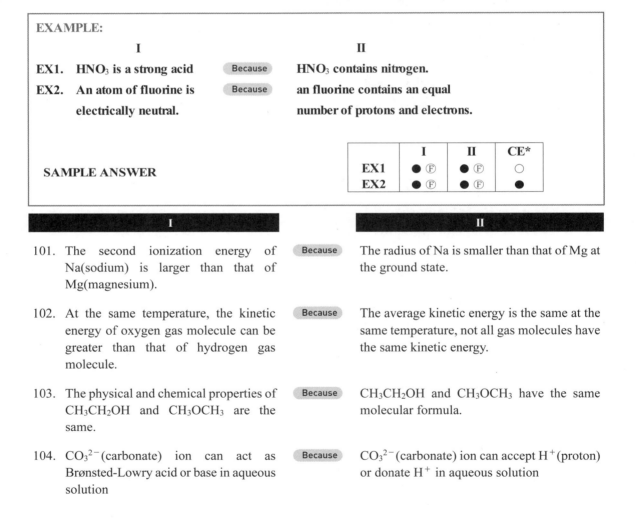

EXAMPLE:

	I		II
EX1.	HNO$_3$ is a strong acid	Because	HNO$_3$ contains nitrogen.
EX2.	An atom of fluorine is electrically neutral.	Because	an fluorine contains an equal number of protons and electrons.

SAMPLE ANSWER

	I	II	CE*
EX1	● F	● F	○
EX2	● F	● F	●

I		II
101. The second ionization energy of Na(sodium) is larger than that of Mg(magnesium).	Because	The radius of Na is smaller than that of Mg at the ground state.
102. At the same temperature, the kinetic energy of oxygen gas molecule can be greater than that of hydrogen gas molecule.	Because	The average kinetic energy is the same at the same temperature, not all gas molecules have the same kinetic energy.
103. The physical and chemical properties of CH$_3$CH$_2$OH and CH$_3$OCH$_3$ are the same.	Because	CH$_3$CH$_2$OH and CH$_3$OCH$_3$ have the same molecular formula.
104. CO$_3^{2-}$(carbonate) ion can act as Brønsted-Lowry acid or base in aqueous solution	Because	CO$_3^{2-}$(carbonate) ion can accept H$^+$(proton) or donate H$^+$ in aqueous solution

105. It is needed to add the limited amount of electrolyte for electrolysis of the water | Because | There are already high concentrations of H^+ and OH^- ion in water.

106. H_3O^+ ion has trigonal planar structure | Because | The oxygen atom in H_3O^+ has a pair of unshared electrons in the Lewis structure

107. The boiling HF is higher than that of HCl | Because | The bond enthalpy of $H-F$ bond is larger than that of $H-Cl$ bond

108. 0.1M of H_2CO_3(aq) is more acidic than that of HNO_3(aq) | Because | At the same concentration of acids, the volume of 0.1M NaOH required to titrate H_2CO_3 is twice that of HNO_3

109. The reaction represented below

$$2NaHCO_3(s)\xrightarrow{\triangle}Na_2CO_3(s)+H_2O(l)$$

$+CO_2(g)$ can be spontaneous at high temperature | Because | The sign of entropy change(ΔS) of endothermic reaction is positive

110. The normal boiling point of water is higher than that of hydrogen fluoride | Because | The polarity of O-H bond in water is larger than that of H-F bond in hydrogen fluoride

111. The volume of 2M NaOH(aq) needed to neutralize 20 mL of 1M H_2SO_4(aq) is greater than that to neutralize of 20 mL of 1M $HC_2H_3O_2$(aq) | Because | H_2SO_4(aq) is strong acid but $HC_2H_3O_2$ is weak acid

112. It is difficult to obtain iron as element in nature, so iron ore is purified to obtain iron. | Because | Iron is a highly reactive metal, which exists mainly in the form of oxides in nature.

113. Protein and nylon are polymers | Because | A polymer is a large molecule, or macromolecule, composed of many repeated subunits

114. When 6mol of H_2(g) and 1mol of N_2(g) react to produce ammonia, the limiting reactant is H_2(g) | Because | The limiting reactant is consumed completely and can determine the theoretical yield in chemical reaction

115. When dry ice sublimates, the temperature of surrounding decreases | Because | The heat energy is absorbed from the surrounding into the system in endothermic reaction.

Part C

Direction: Each of the questions or incomplete statements below is followed by five suggested answer or completions. Select the one that is best in each case and then fill in the corresponding circle on the answer sheet.

26. The below are the enthalpy changes of several reactions

$2C(s, graphite) + 3H_2(g) + 1/2O_2(g) \rightarrow C_2H_5OH(l)$
$\qquad \Delta H = -278kJ/mol$
$C(s, graphite) + O_2(g) \rightarrow CO_2(g)$
$\qquad \Delta H = -394kJ/mol$
$H_2(g) + 1/2O_2(g) \rightarrow H_2O(l)$
$\qquad \Delta H = -286kJ/mol$

Which of the following is correct for the heat of reaction represented below?

$C_2H_5OH(l) + 3O_2(g) \rightarrow 2CO_2(g) + 3H_2O(l)$

(A) –208kJ/mol   (B) –384kJ/mol
(C) –958kJ/mol   (D) –1368kJ/mol
(E) –1924kJ/mol

27. Which of the following is not correct for the reaction represented below?

$...Pb(NO_3)_2(s) \rightarrow ...PbO(s) + ...NO_2(g) + ...O_2(g)$

(A) The sum of all coefficients as whole number is 9
(B) when the forward reaction occurs the entropy increases
(C) when 1mol of $Pb(NO_3)_2(s)$ are decomposed, the volume of $NO_2(g)$ produced is 44.8L at STP ($0\,^\circ\!C$, 1atm)
(D) when 24g of oxygen gas is obtained from decomposition of 2mol of $Pb(NO_3)_2$, the % yield is 75%
(E) when 1mol of $NO_2(g)$ formed 2mol of electrons

are transferred.

28. The below is the unbalanced redox reaction in acidic condition

$...SO_3^{2-}(aq) + ...MnO_4^-(aq)$
$\rightarrow ...SO_4^{2-}(aq) + ...Mn^{2+}(aq)$

Which of the following is(are) correct?

I. $SO_3^{2-}(aq)$ is reducing agent
II. The oxidation number of Mn in MnO_4^- is $+7$
III. The mol ratio of SO_3^{2-} : MnO_4^- is 2 : 5 in balanced equation

(A) I only   (B) II only
(C) I, and II only   (D) II, and III only
(E) I, II, and III

29. Which of the following is correct for the solubility of $Zn(OH)_2(s)$?

$Zn(OH)_2(s) \rightleftharpoons Zn^{2+}(aq) + 2OH^-(aq)$

I. $Zn(OH)_2$ is partially dissociated in water
II. When adding the $NaOH(s)$ in $Zn(OH)_2(aq)$, the solubility of $Zn(OH)_2(s)$ increase
III. The solubility of $Zn(OH)_2(s)$ in pure water is the same as the concentration of $OH^-(aq)$

(A) I only   (B) II only
(C) I, and II only   (D) II, and III only
(E) I, II, and III

30. Which of the following is correct for the properties of gas?

(A) The density of $H_2(g)$ with 1atm, 100K is the same that of $O_2(g)$ with 0.5atm 400K

(B) The average kinetic energy is the same regardless of kinds of molecules and pressure at the same temperature.

(C) When mixing two gases, the fraction of each gas is equal to the ratio of the pressure of each gas before mixing.

(D) When 3L of 4atm $N_2(g)$ and 2L of 3atm $O_2(g)$ are mixed in rigid container the mol fraction of $N_2(g)$ is 0.5

(E) The rate of effusion of gas is directly proportional to the temperature.

31. Which of the following reaction shows the greatest increase in entropy change for the forward reaction?

(A) $C_3H_8(g) + 5O_2(g) \rightarrow 2CO_2(g) + 4H_2O(l)$

(B) $4Al(s) + 3O_2(g) \rightarrow 2Al_2O_3(g)$

(C) $Zn(s) + 2HCl(aq) \rightarrow ZnCl_2(aq) + H_2(g)$

(D) $NH_4OCONH_2(s) \rightarrow 2NH_3(g) + CO_2(g)$

(E) $CO_2(g) + 2NH_3(aq) \rightarrow NH_2COO^-(aq) + NH_4^+(aq)$

32. Which of the following is not correct for the represented unbalanced reaction?

$$...NO(g) + ...Cl_2(g) + ...H_2O(l) \rightarrow ..NO_3^-(aq) + ...Cl^-(aq) + ...H^+(aq)$$

(A) The reaction occurs at acidic condition

(B) When the equation is balanced as whole number the sum of all coefficient is 25

(C) $Cl_2(g)$ is oxidizing agent

(D) The mole number ratio of transferred electrons, $NO(g) : Cl_2(g)$ is 2 : 3

(E) The compound with the largest positive oxidation number is $NO_3^-(aq)$

33. When 2M of $Na_2CO_3(aq)$ 400mL and 3M of $Ca(NO_3)_2(aq)$ 600mL are mixed to form a 1L of solution, what is the molarity of $Ca^{2+}(aq)$?

(A) 0.5M   (B) 1.0M

(C) 1.5M   (D) 2.0M

(E) 3.0M

34. The below is the oxygen evolving reaction

$$...KClO_3(s) \xrightarrow[\triangle]{MnO_2(s)} ...KCl(s) + ...O_2(g)$$

Oxygen gas can be collecting from water displacement. The volume of gas collected 500mL at 25℃, 750mmHg, the water line of inside and outside of collecting bottle is the same. The vapor pressure of water at 25℃ is 23.8mmHg. Which of the following is not correct for the reaction?

(A) The decomposition of $KClO_3(s)$ is oxidation–reduction reaction.

(B) $MnO_2(s)$ is consumed as oxidizing agent.

(C) When the reaction above is balanced, the ratio of coefficient, $O_2/KClO_3$ is 1.5.

(D) When the forward reaction occurs, the entropy of system increases.

(E) If the actual temperature is higher than 25°C, the mass of reacting $KClO_3$ calculated is greater than the actual value.

35. The mole number of carbon as reducing agent required to obtain 4 mol of iron from oxide of iron is 3 mol. Which of the following is correct formula of iron oxide when carbon is completely oxidized?

(A) FeO   (B) FeO_2 (C) Fe_2O_3

(D) Fe_3O_4 (D) Fe_4O_3

36. Which of the following is not correct for the reaction represented below?

$$KCl(s) + 19kJ/mol \xrightarrow{H_2O\,(l)} KCl(aq)$$

(A) When the solute is dissolved in water the temperature of solution decrease.

(B) The solubility of KCl(s) increase as the temperature increase.

(C) The dissolving process is endothermic.

(D) Solubility can be increased by adding KCl (s) to the saturated aqueous solution and rapidly stirring.

(E) The reason why KCl (aq) solution can be formed is the effect of increasing of entropy in dissolving of KCl(s)

37. The below is the ionization reaction of ethanoic acid, $HC_2H_3O_2$, and ionization constant, Ka

$$HC_2H_3O_2(aq) + H_2O(l) \rightleftharpoons H_3O^+(aq) + C_2H_3O_2^-(aq)$$
$$Ka = 1.8 \times 10^{-5}$$

Which of the following is not correct for the reaction?

(A) Ethanoic acid can be partial ionized in water

(B) Brønsted–Lowry acids are $HC_2H_3O_2$(aq) and H_3O^+(aq)

(C) $C_2H_3O_2^-$(aq) is stronger base than H_2O(l)

(D) At equilibrium state, the concentration of $[HC_2H_3O_2]$ and $[C_2H_3O_2^-]$ are the same.

(E) Net ionic equation of neutralization reaction of NaOH(aq) and ethanoic acid is $HC_2H_3O_2$(aq) + OH^-(aq)→H_2O(l) + $C_2H_3O_2^-$(aq)

38. When 0.2L of 0.5M of Na_2SO_4(aq) is added to 0.4L of $AgNO_3$(aq), there are only two kinds of ions in mixture. What is the mol number of $AgNO_3$(aq) initially?

(A) 0.05mol   (B) 0.1mol   (C) 0.2mol

(D) 0.4mol   (E) 0.8mol

39. 800mL of 0.1M NaOH (aq) was consumed to neutralize 500mL of HCN (aq) completely. Which of the following is not correct for the titration? (the initial and final temperature are the same as 25℃. Ka of HCN is 5×10^{-10})

(A) The concentration of HCN(aq) is 0.16M

(B) The pH at endpoint is higher than 7

(C) $[Na^+]$ is larger than $[CN^-]$ at equivalence point

(D) The volume of 0.1 M of NaOH (aq) required to completely neutralize the same concentration and volume of HCl (aq) as HCN is greater than in HCN.

(E) CN^-(aq) is stronger base than H_2O(l)

40. When the Helium gas and oxygen gas are mixed in evacuate rigid container with 11.2L at 0℃, the partial pressure of oxygen gas is 0.2atm and total pressure is 1.0atm. which of the following is not correct for the mixture of gases?

(A) The mole ratio, He : O_2 is 4 : 1

(B) The total mole number of mixture is 0.5mol

(C) The volume of $O_2(g)$ at 0℃, 1atm is 2.24L

(D) The volume of He(g) at 0℃, 2atm is 4.48L

(E) The ratio of density, He : O_2 is 1: 1

41. Which of the following is not correct for the properties of solution?(molar lowering constant of water $k_{f=}-1.86$℃/m)

(A) Hydration is the solvation of water

(B) Water and hexane are immiscible

(C) The vapor pressure depression of solution which contains nonvolatile and nonelectrolytic solute directly proportional to the mole fraction of solvent

(D) The osmotic pressure of glucose solution, $C_6H_{12}O_6(aq)$ is smaller than that of sodium chloride solution, NaCl(aq) at same molarity.

(E) The freezing point lowering of 1m of $CaCl_2(aq)$ is –5.58℃, if $CaCl_2$ can be ionized completely in water

42. When 100mL of 1M $Na_2SO_4(aq)$ is mixed with 100mL of 2M $Ca(NO_3)_2(aq)$, which of the following is not correct for the reaction?

(A) The net ionic equation is $Ca^{2+}(aq)+SO_4^{2-}(aq)$ → $CaSO_4(s)$.

(B) The only color of flame reaction of solution is yellow.

(C) $NO_3^-(aq)$ is the greatest concentration in mixture.

(D) The total mole number of particles is mixture is 0.7mol.

(E) When $CaCl_2(aq)$ is added to the mixture, additional precipitation can not be observed.

43. Which of the following is correct for the crystalline structure?

(A) Atomic crystalline can not conduct electricity.

(B) Metallic crystalline is formed by electrostatic force between nucleus and free electrons

(C) Quartz and glass are isomer relationship.

(D) Molecular crystalline is formed by network covalent bond.

(E) Ionic crystalline is formed by coulomb force between cation and anion

$$C(s, graphite)+O_2(g) \rightarrow CO_2(g) \quad \Delta H = -393.5kJ$$
$$2CO(g)+O_2(g) \rightarrow 2CO_2(g) \quad \Delta H = -566.0kJ$$

44. What is the enthalpy change(ΔH) of the reaction represented below?

$$C(s, graphite)+1/2O_2(g) \rightarrow CO(g)$$

(A) –110.5kJ   (B) –172.5kJ   (C) –479.5kJ
(D) –762.5kJ   (E) –959.5kJ

45. How many moles of Cu(s) are formed when 2moles of $CH_5N(g)$ are reacted with CuO(s) completely?

$$CH_5N(g) +...CuO(s)$$
$$\rightarrow ...CO_2(g) +...N_2(g) +...Cu(s)+...H_2O(l)$$

(A) 1mol   (B) 3mol   (C) 5mol
(D) 7mol   (E) 9mol

46. Effusion rate of unknown gas A(g) is 2 times as fast as Ar(g)(molar mass : 40g/mol) at same temperature and pressure. Which of the following is correct for the equation of density of A(g) at $0°C$, 1atm?

(A) $\frac{40}{22.4} \times \sqrt{2}$ (g/L)

(B) $\frac{10}{22.4} \times \frac{1}{\sqrt{2}}$ (g/L)

(C) $\frac{40}{22.4} \times \frac{1}{\sqrt{2}}$ (g/L)

(D) $\frac{40}{22.4} \times \frac{1}{4}$ (g/L)

(E) $\frac{40}{22.4} \times 4$ (g/L)

47. 2mol of $H_2(g)$ and 3mol of He(g) are contained in rigid container. the partial pressure of $H_2(g)$ is 400mmHg. What is the total pressure of container?

(A) 600mmHg   (B) 700mmHg
(C) 800mmHg   (D) 900mmHg
(E) 1000mmHg

48. The below reaction important in production of nitrogen fertilizer occurs spontaneously at low temperature.

$$2NH_3(g) + CO_2(g) \rightarrow NH_2CONH_2(aq) + H_2O(l)$$

Which of the following is correct for the reaction represented above?

(A) When the forward reaction proceeds, the temperature of surrounding decreases.
(B) When forward reaction proceeds, the entropy of system increases.

(C) The reverse reaction is always spontaneous regardless of the temperature.
(D) $CO_2(g)$ is reducing agent because oxidation number of carbon increases in the forward reaction.
(E) The state of products are more stable than that of reactants.

49. Which of the following is not correct for the conjugate acid and base pair?

(A) $H_2PO_4^-$ and HPO_4^{2-}
(B) H_2O and H_3O^+
(C) NaOH and Na^+
(D) HCO_3^- and CO_3^{2-}
(E) H_3PO_3 and $H_2PO_3^-$

50. When the reaction represented below is balanced, when 2mole of $SiCl_4(l)$ react with excess of $NH_3(g)$, what is the amount of $NH_4Cl(s)$ produced?

$$...SiCl_4(l) + ...NH_3(g) \rightarrow$$
$$...Si(NH)_2(s) + ...NH_4Cl(s)$$

(A) 2mol   (B) 4mol   (C) 6mol
(D) 8mol   (E) 10mol

51. Which of the following electron transitions is related with the shortest wavelength light emission of hydrogen atom?

(A) 2s→2p   (B) 1s→3s   (C) 2p→1s
(D) 4p→2s   (E) 5d→2p

52. Which of the following is correct for the electrochemical cells?

(A) Cathode is (–) electrode in spontaneous electrochemical cell.

(B) Anode is (+) electrode in electroplating of metal.

(C) The mass of anode increases when $Cu(NO_3)_2$(aq) electrolyzed.

(D) NaCl(aq) can be used in electrolysis of water.

(E) 1.2×10^{24} electrons are needed to produce 1mol of Cu(s) at anode in electrolysis of $CuSO_4$(aq).

53. The boiling of which of the followings is highest at 1atm?(glucose : 180g/mol, NaCl : 58.5g/mol, Na_2SO_4 : 142g/mol, $Ca(NO_3)_2$: 164g/mol)

(A) A solution of 30.0g of glucose dissolved in 200g of water.

(B) A solution of 11.7g of sodium chloride dissolved in 500g of solution.

(C) A solution of 5.68g of sodium sulfate dissolved in 200g of water.

(D) A solution of 82.0g of calcium nitrate dissolved in 2000g of water.

(E) A mixture of solution (C) and (D)

The solutions represented below are mixtures of acid and base.

Solution A : 0.1M of H_2SO_4(aq) 15mL + 0.15M of $Ca(OH)_2$(aq) 10mL

Solution B : 0.3M of $HC_2H_3O_2$(aq) 30mL + 0.2M of NaOH(aq) 20mL

Solution C : 0.1M of HCl(aq) 40mL + 0.4M of NH_3(aq) 10mL

54. Which of the following is correct for the solutions listed above?

(A) The concentration of $[H^+]$ in solution C is 10^{-7}M

(B) The electric conductivity of solution A is the largest.

(C) The pH of solution A is highest at 298K.

(D) The ion product of water, $Kw = [H^+][OH^-]$ is highest in solution B at 298K.

(E) The solution C can resist to change in pH when limited amount of strong acid or strong base are added.

55. 30℃ of 20g metal A and 40℃ of 10g of metal B is added to 100g of 80℃ water in each. the final temperature of A is 50℃ and B is 60℃. Which of the following is correct for the ratio of specific heat capacity, A : B ?

(A) 1 : 1   (B) 1 : 2   (C) 2 : 1

(D) 3 : 4   (E) 4 : 3

56. When the reaction is balanced in whole number ratio, which of the following is not correct for the reaction below?

$$...Ag(s) + ...O_2(g) + ...H_2S(g) \rightarrow$$
$$...Ag_2S(s) + ...H_2O(l)$$

(A) Ag(s) is reducing agent.

(B) Oxidation number changes of O(oxygen) and S(sulfur) are the same

(C) The oxidation number of Ag(s) changes from 0 to +1

(D) The sum of reaction coefficients of reactants is larger than that of products.

(E) 4mol of electrons are transferred from reducing agent to oxidizing agent when 1mol of oxygen molecules are reacted completely.

57. 20.0mL of 0.020mol $\cdot$ L^{-1} H$_2$SO$_4$(aq) is added to 30.0mL of 0.040mol $\cdot$ L^{-1} Ca(OH)$_2$(aq). Which of the following is not correct for this reaction at 25℃?

(A) The reaction is neutralization reaction.

(B) Final pH of the mixture is larger than 7

(C) The concentration of Ca^{2+}(aq) is 0.016M

(D) The mole number of water produced from the reaction is 0.0008mol.

(E) The solubility of the salt from the reaction represented above is increasing as increasing of pH at constant temperature.

58. Which of the following is correct for the molar concentration of Na$^+$(aq) when 11.7g of NaCl(s) is dissolved in water completely to make 2L of solution?(molar mass of NaCl : 58.5g/mol)

(A) 0.1M   (B) 0.2M   (C) 0.3M

(D) 0.4M   (E) 0.5M

59. Which of the followings is the most acidic when 0.01mol of each acid is dissolved in 0.1L of water?

Acid	pKa$_1$	pKa$_2$
Acetic acid	4.74	
Lactic acid	3.86	
Hydrocyanic acid	9.21	
Benzoic acid	4.19	
Carbonic acid	6.37	10.3

(A) Acetic acid   (B) Lactic acid

(C) Cyanic acid   (D) Benzoic acid

(E) Carbonic acid

60. 8atm, 1L of NO(g) is mixed with 4atm 2L of O$_2$(g) in 4L of rigid container. Which of the following is correct for the complete combustion reaction of NO(g) with O$_2$(g) to form dinitrogen pentoxide gas?(the temperature remains constant)

(A) When the reaction is balanced, the sum of reaction coefficient as the whole number is 7.

(B) Before the combustion the mole fraction oxygen gas is larger than that of nitrogen monoxide gas in mixture.

(C) O$_2$(g) is limiting reactant.

(D) If the partial pressure of product is 0.8atm, the percent yield is 60%

(E) The total pressure is 1.5atm after the reaction.

61. When 0.01 mol of hydrocarbon, C$_x$H$_y$, is completely burned, 1.76g of CO$_2$ and 0.9g of H$_2$O are produced. Which of the following is correct?(The atomic mass of H : 1g/mol, C : 12g/mol, O : 16g/mol)

I. Molecular formula is C$_2$H$_5$

II. 145.6L of oxygen gas is required for the complete combustion of 1mol of hydrocarbon at STP(0℃, 1atm)

III. The density of hydrocarbon is larger than that of propane(C$_3$H$_8$) at same temperature and pressure

(A) I only   (B) II only

(C) I and II only   (D) II and III only

(E) I and III only

62. $H_2S(g) + I_2(s) \rightleftharpoons 2HI(g) + S(s)$

    $K_p = 1.00 \times 10^{-5}(60℃)$

Which of the following is correct for the reaction above?

(A) The reverse reaction is exothermic.

(B) The reaction is spontaneous at 60℃

(C) K_p value of $2H_2S(g) + 2I_2(s) \rightleftharpoons 4HI(g) + 2S(s)$ is $2.00 \times 10^{-5}(60℃)$

(D) When S(s) is added to equilibrium state, the reaction goes to reverse and reaches to new equilibrium state.

(E) When the external pressure is added to equilibrium state and the volume of container is reduced, the reaction goes to forward, and reaches to new equilibrium state.

63. Which of the following is the least useful for determining the concentration of 0.2L of NaOH(aq) through the titration with 0.1M of $HNO_3(aq)$?

(A) Buret   (B) Indicator

(C) Thermometer   (D) Pestle and mortar

(E) Erlenmeyer flask

64. When the 1.00M of excess of hydrochloric acid is added to 10g of calcium carbonate at 25℃, 1atm Which of the following is correct?(the molar mass of calcium carbonate is $100g \cdot mol^{-1}$)

(A) The reaction is oxidation-reduction reaction

(B) When the produced gas is added to $Ca(OH)_2(aq)$, calcium carbonate precipitate is formed.

(C) The volume of produced gas is $2.24 \times \dfrac{273}{298}L$

(D) When the 2.00M of excess of hydrochloric acid is used, the total volume of gas produced increase.

(E) The produced gas is base anhydride.

65. $...C_8H_8O_3(s) + ...NaOH(aq) + H_2SO_4(aq)$
    $\rightarrow ...(A) + 2CH_3OH(aq) + Na_2SO_4(aq)$

The above represented is the chemical reaction for synthesis of salicylic acid. Which of the following is correct for the (A)?

(A) $C_7H_6O_2$ (B) $C_7H_5O_3$ (C) $C_7H_6O_3$

(D) $C_7H_7O_2$ (E) $C_7H_7O_3$

66. The below is the dinitrogen trioxide decomposition reaction.

$$...N_2O_3(g) \rightleftharpoons ...NO(g) + ...O_2(g)$$

$N_2O_3(g)$ is added to evacuated 1.0L rigid container and initial pressure of $N_2O_3(g)$ is 0.6atm At the equilibrium state, the partial pressure of $O_2(g)$ is 0.1atm. Which of the following is not correct for this reaction?(temperature remains constant)

(A) The activation energy of forward reaction is larger than that of reverse reaction.

(B) The rate of reverse reaction rate decrease as the temperature decreases.

(C) If container volume is decreased. the reaction go shift to the reverse reaction

(D) The K value increases as the temperature increases.

(E) The forward reaction is spontaneous at experimental temperature

67. The below is the table of indicator for titration of acid and base.

Indicator	pH range	color	
		Acidic	Basic
Methyl orange	3.1-4.4	orange	yellow
Bromphenol blue	2.8-4.6	yellow	blue
Methyl red	4.2-6.3	red	yellow
Phenolphthalein	8.3-10.0	colorless	red

When 2000mL of HNO_2(nitrous acid) is titrated with 0.5M of NaOH(aq), the volume of titrant is 400mL at endpoint. Which of the following is not correct for the titration? ($Ka(HNO_2) = 4.5 \times 10^{-4}$)

(A) Phenolphthalein is suitable indicator for confirm the endpoint.

(B) The concentration of titrand is 0.1M

(C) The initial pH of HNO_2(aq) before the titration is higher than 1

(D) Na^+(aq) is the highest concentration ion at equivalence point.

(E) When all conditions are the same and acid is changed to HCl, the volume of NaOH (aq) added to neutralize the acid is larger than 400mL.

68. Which of the following is not correct for the 3 phases ; solid, liquid and gas?

(A) When pressure is applied to gas, it can be condensed to liquid.

(B) At the constant temperature, the pressure needed to condensed the He(g) is larger than that of Ar(g)

(C) There is no empty space between liquid molecules, but in the case of gas.

(D) The kinetic energy of gas particles in a container of constant temperature and pressure are not all the same.

(E) Solid particles vibrate at fixed position and have no fluidity.

69. The below is the vapor pressure of some liquid

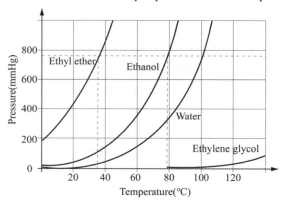

Which of the following is not correct for the vapor pressure?

(A) The vapor pressure of each substance is not the same at it's normal boiling point.

(B) The vapor pressure increase as the temperature increase.

(C) All the point at the each curve represent the equilibrium state between liquid and vapor

(D) The boiling point of liquid increases as the intermolecular force increases.

(E) 120℃ water can be exist at higher external pressure

70. The below is the standard reduction potential of substances.

Half reaction	$E^{o}_{red}(V)$
$Cu^{2+}(aq)+2e^{-}\rightarrow Cu(s)$	$+0.34$
$Ag^{+}(aq)+e^{-}\rightarrow Ag(s)$	$+0.80$
$Zn^{2+}(aq)+2e^{-}\rightarrow Zn(s)$	-0.76
$Cl_2(g)+2e^{-}\rightarrow 2Cl^{-}(aq)$	$+1.30$

Which of the following is not correct for the redox reaction?

(A) Cl_2 is the strongest oxidizing agent

(B) $Zn(s)$ is the strongest reducing agent

(C) The standard cell potential $Cu^{2+}(aq)+Zn(s)\rightarrow$ $Cu(s)+Zn^{2+}(aq)$ is $+1.10V$

(D) When electric current is applied to $CuCl_2(aq)$ $Cu(s)$ metal can be deposited in anode

(E) The standard cell potential $2Ag^{+}(aq)+Zn(s)\rightarrow$ $2Ag(s)+Zn^{2+}(aq)$ is $+1.56V$

MEMO

Answer Key

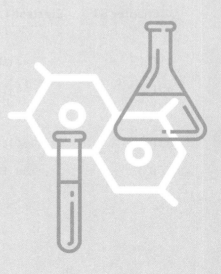

Chapter Review Test

Chapter 01	Matter and Measurement

01 (C)	02 (A)	03 (B)	04 (D)
05 TTCE	06 FT	07 TTCE	08 (D)
09 (C)	10 (A)	11 (D)	12 (C)
13 (D)	14 (C)	15 (B)	16 (C)
17 (B)	18 (B)	19 (D)	20 (E)

Chapter 02	Atom, Molecule and Ion

21 (C)	22 (E)	23 TT	24 TF
25 (E)	26 (C)	27 (A)	28 (E)
29 (B)	30 (D)	31 (C)	32 (B)
33 (D)	34 (C)	35 (C)	36 (E)
37 (B)	38 (B)	39 (E)	40 (E)

Chapter 03	Electron Configuration

41 (E)	42 (D)	43 (C)	44 (B)
45 (A)	46 (C)	47 (E)	48 (B)
49 TT	50 FT	51 (D)	52 (D)
53 (C)	54 (C)	55 (D)	56 (C)
57 (C)	58 (B)	59 (B)	60 (C)

Chapter 04	**Periodic Table and Periodicity**		
61 (D)	62 (E)	63 (B)	64 (D)
65 FT	66 TF	67 FF	68 (E)
69 (D)	70 (E)	71 (D)	72 (E)
73 (C)	74 (C)	75 (B)	76 (B)
77 (B)	78 (E)	79 (C)	80 (B)

Chapter 05	**Chemical Bond**		
81 (E)	82 (C)	83 (D)	84 TTCE
85 TTCE	86 TTCE	87 (D)	88 (B)
89 (D)	90 (E)	91 (D)	92 (B)
93 (D)	94 (D)	95 (E)	96 (C)
97 (A)	98 (E)	99 (E)	100 (D)

Chapter 06	**Gas**		
101 (E)	102 (D)	103 FT	104 TT
105 TTCE	106 TT	107 TF	108 (E)
109 (E)	110 (D)	111 (E)	112 (C)
113 (B)	114 (E)	115 (D)	116 (D)
117 (B)	118 (A)	119 (B)	120 (C)

Chapter 07	Liquid, Solid, and Phase change		
121 (E)	122 (B)	123 (B)	124 (C)
125 FT	126 TT	127 TTCE	128 (C)
129 (D)	130 (A)	131 (D)	132 (D)
133 (E)	134 (B)	135 (B)	136 (C)
137 (B)	138 (C)	139 (D)	140 (E)

Chapter 08	Solution		
141 (E)	142 (A)	143 (D)	144 TTCE
145 TTCE	146 FT	147 TF	148 (B)
149 (C)	150 (D)	151 (D)	152 (C)
153 (B)	154 (D)	155 (B)	156 (D)
157 (A)	158 (B)	159 (E)	160 (C)

Chapter 09	Thermodynamics		
161 (C)	162 (B)	163 (A)	164 (C)
165 (B)	166 (C)	167 (E)	168 (E)
169 (B)	170 (E)	171 (D)	172 (A)
173 (C)	174 (E)	175 (E)	176 (D)
177 (D)	178 (B)	179 (D)	180 (D)

Chapter 10	Chemical Kinetics		
181 (E)	182 (C)	183 TTCE	184 (A)
185 (A)	186 (E)	187 (A)	188 (D)
189 (E)	190 (C)	191 (B)	192 (C)
193 (C)	194 (C)	195 (C)	196 (E)
197 (C)	198 (B)	199 (C)	200 (E)

Chapter 11	Chemical Equilibrium		
201 (B)	202 (A)	203 TF	204 FT
205 TF	206 (E)	207 (A)	208 (D)
209 (D)	210 (D)	211 (E)	212 (D)
213 (A)	214 (A)	215 (D)	216 (B)
217 (D)	218 (D)	219 (C)	220 (B)

Chapter 12	Acid and Base		
221 (B)	222 (A)	223 FT	224 TTCE
225 (D)	226 (D)	227 (D)	228 (C)
229 (A)	230 (D)	231 (D)	232 (C)
233 (D)	234 (B)	235 (D)	236 (C)
237 (C)	238 (B)	239 (A)	240 (D)

Chapter 13	Electrochemistry		
241 (C)	242 (B)	243 TTCE	244 TTCE
245 (C)	246 (A)	247 (E)	248 (E)
249 (D)	250 (D)	251 (C)	252 (D)
253 (D)	254 (D)	255 (C)	256 (A)
257 (D)	258 (D)	259 (C)	260 (D)

Chapter 14	Organic Chemistry		
261 (D)	262 (C)	263 (E)	264 (A)
265 (D)	266 (C)	267 (E)	268 (A)
269 (D)	270 (B)		

Chapter 15	Nuclear Chemistry		
271 FT	272 FT	273 FT	274 TT
275 (B)	276 (C)	277 (C)	278 (A)
279 (B)	280 (B)		

Final Full Test

Full Test No.1

[Part A]

1 (A)

   CO_2 is nonpolar molecule.

2 (D)

   Ionic bond is formed through electrostatic force between cation and anion.

3 (E)

   Hydrogen bond is strong intermolecular force.

4 (A)

   Noble gas is stable because of it's stable electron configuration.

5 (E)

   The electron configuration of Na^+ is the same as Ne.

6 (C)

   The lower left element in the periodic table, the smaller the first ionization energy.

7 (B)

   The radius of an isoelectronic ion increases as the atomic number decreases.

8 (B)

   Carbonate reacts with acid to produce carbon dioxide.

9 (A)

   $CuI(s)$ is white precipitate.

10 (C)

   $PbS(s)$ is balck precipitate.

11 (E)

   The color of the flame reaction of calcium is orange.

12 (C)

13 (B)

14 (D)

15 (A)

16 (D)

In the transition metal of the d-block, the d orbital is partially filled with electrons.

17 (E)

In the Lanthanides, the f orbital is partially filled with electrons.

18 (B)

Silicon is a raw material for semiconductors.

19 (E)

20 (D)

21 (B)

22 (E)

23 (A)

24 (A)

25 (D)

[Part B]

101 FF

102 TTCE

103 TTCE

104 TF

105 FT

106 TTCE

107 FT

108 FF

109 TF

110 FT

111 TT

112 TT

113 TTCE

114 FF

115 TF

[Part C]

26 (C)

27 (B)

The radius of the isoelectronic decreases as the atomic number increases.

28 (C)

29 (C)

Average speed of molecule is proportional to $\sqrt{\dfrac{T}{M}}$.

30 (D)

Generally, halogens are good oxidizing agent in chemical reaction.

31 (A)

32 (D)

The sum of bond energy of reactants is larger than that of products in endothermic reaction.

33 (E)

34 (C)

35 (B)

36 (D)

37 (C)

38 (B)

39 (C)

40 (A)

41 (C)

The proton acceptor in the reaction is called the Brønsted-Lowry base.

42 (C)

43 (C)

44 (D)

45 (A)

46 (B)

47 (D)

48 (E)

49 (C)

50 (B)

51 (C)

Temperature is intensive property.

52 (C)

53 (B)

54 (B)

55 (B)

56 (D)

57 (E)

The half-life of first order reaction is constant.

58 (B)

59 (B)

60 (E)

61 (B)

62 (E)

63 (C)

64 (E)

65 (D)

66 (C)

67 (D)

68 (D)

69 (A)

70 (C)

Full Test No.2

[Part A]

1 (A)

2 (E)

3 (C)

4 (D)

5 (A)

6 (C)

7 (C)

8 (A)

9 (D)

10 (C)

11 (D)

12 (A)

13 (E)

Cations of transition metals can exhibit color in aqueous solutions.

14 (B)

15 (E)

Even if it has a polar covalent bond, it is a non-polar molecule if its molecular structure is symmetric.

16 (C)

17 (E)

18 (B)

19 (A)

20 (A)

21 (E)

22 (C)

23 (E)

24 (B)

25 (A)

[Part B]

101 TT

102 TF

103 TTCE

104 FF

105 FT

106 TF

As the atomic number increases, the atomic radius decreases in the same period.

107 FT

108 TTCE

109 TF

As the temperature increases, the reaction rate increases.

110 FT

111 TTCE

112 TF

113 TF

114 FT

The stable electron configuration of ground state satisfies the Aufbau principle.

115 TF

[Part C]

26 (A)

27 (E)

28 (E)

29 (E)

30 (A)

31 (B)

The unit of molar concentration is mol/L and the number of moles is the product of the molar concentration and volume of the solution.

32 (C)

33 (E)

34 (D)

35 (D)

36 (E)

When cations are produced from the period 4 transition metals, the electrons of the 4s orbital are first removed.

37 (D)

38 (E)

39 (D)

40 (D)

41 (D)

42 (B)

43 (E)

44 (E)

Reaction (E) is a Lewis acid-base reaction.

45 (D)

46 (D)

47 (C)

48 (A)

49 (E)

50 (B)

51 (E)

52 (B)

53 (C)

54 (C)

55 (D)

56 (C)

57 (D)

58 (D)

59 (D)

60 (A)

61 (B)

62 (D)

63 (A)

64 (E)

65 (C)

66 (C)

67 (B)

68 (C)

69 (C)

70 (E)

Full Test No.3

[Part A]

1 (A)

Oxides of sulfur and nitrogen are substances that cause acid rain

2 (B)

Carbon(graphite) has one free electron per atom.

3 (D)

4 (E)

5 (C)

6 (E)

7 (C)

8 (D)

9 (B)

10 (D)

11 (C)

12 (A)

13 (E)

14 (C)

15 (D)

16 (E)

17 (B)

18 (A)

19 (B)

20 (D)

21 (B)

22 (E)

23 (C)

24 (D)

25 (A)

[Part B]

101 TF

102 TTCE

103 FT

104 TTCE

105 TT

106 FT

107 TTCE

108 FT

109 FT

At equilibrium, the rate of forward reaction rate is the same as reverse reaction.

110 FT

111 TF

112 FT

The exothermic reaction with decreasing entropy is spontaneous at low temperatures.

113 TTCE

114 TTCE

115 TTCE

[Part C]

26 (C)

27 (D)

In OF_2, the oxidation number of oxygen is +2.

28 (C)

When an electron is transferred, the wavelength of light absorbed or emitted is inversely proportional to the energy difference.

29 (B)

30 (D)

31 (B)

32 (D)

33 (C)

34 (C)

It is a physical change that a solute dissolves in a solvent.

35 (E)

Pure solids and liquids are not included in the equilibrium constant.

36 (B)

37 (D)

38 (E)

39 (D)

40 (B)

41 (C)

The oxidation number of Cr increased from +6 to +3, and 1 mol of $Cr_2O_7^{2-}$ contains 2mol of Cr^{3+}.

42 (E)

43 (C)

44 (B)

45 (D)

46 (E)

47 (D)

Carbonate reacts with acid to produce $CO_2(g)$

48 (E)

49 (A)

The colligative properties is the properties that depends only on the number of particles regardless of the type of solute.

50 (B)

51 (B)

52 (D)

53 (D)

54 (D)

55 (B)

56 (B)

57 (B)

58 (B)

59 (D)

60 (D)

61 (D)

62 (A)

63 (B)

64 (C)

65 (B)

66 (C)

Amphoteric substances are substances that can react with both acids or bases.

67 (B)

Salts resulting from neutralization of strong bases and weak acids exhibit basicity.

68 (A)

69 (D)

70 (A)

Full Test No.4

[Part A]

1 (C)

2 (B)

3 (A)

4 (E)

5 (A)

6 (E)

7 (B)

8 (D)

9 (B)

10 (B)

11 (D)

12 (E)

13 (C)

14 (D)

15 (E)

16 (B)

17 (B)

18 (A)

19 (E)

20 (D)

21 (B)

22 (A)

23 (B)

24 (D)

25 (E)

[part B]

101 TTCE

102 FT

103 FT

104 TTCE

105 TT

106 TT

107 TT

108 FT

109 FF

110 FT

111 TTCE

112 TT

113 TTCE

Chromatography is a method of separating a mixture by using the difference in the affinity of a component to a mobile phase and a stationary phase.

114 FF

115 FT

Water displacement is a method of collecting gases with low solubility in water.

[Part C]

26 (B)

27 (D)

28 (C)

29 (E)

30 (B)

31 (B)

32 (D)

33 (D)

34 (D)

35 (C)

The ratio of the partial pressure of gas in the same container is proportional to the ratio of the number of moles.

36 (D)

37 (C)

38 (C)

39 (C)

The indicator is added in small amounts to the substance to be titrated.

40 (C)

41 (E)

42 (B)

43 (D)

44 (D)

45 (C)

46 (B)

47 (C)

48 (C)

The electron configuration of Fe^{3+} is $[Ar]3d^5$.

49 (B)

50 (C)

The cell potential is an intensity property and is not affected by the amount of substance

51 (D)

52 (E)

53 (C)

54 (C)

55 (C)

56 (C)

57 (A)

58 (C)

59 (D)

The main reason of the difference of intermolecular force between HCl and HBr is the LDF.

60 (E)

61 (D)

62 (E)

63 (C)

64 (C)

65 (C)

66 (E)

67 (D)

68 (B)

69 (A)

70 (C)

Full Test No.5

[Part A]

1 (D)

2 (B)

3 (E)

4 (C)

5 (C)

6 (D)

7 (A)

8 (B)

9 (A)

10 (B)

11 (B)

12 (E)

13 (D)

14 (D)

15 (A)

16 (B)

17 (E)

18 (D)

19 (B)

20 (D)

21 (C)

22 (D)

23 (E)

24 (C)

25 (A)

[Part B]

101 FT

102 TTCE

103 FT

Structural isomers have different physical and chemical properties.

104 FF

105 TF

106 FT

107 TT

108 FT

109 TF

110 TF

The polarity of the bonds is larger in O–F than in O–H, but the number of hydrogen bonds per molecule is greater in H_2O than in HF.

111 TT

H_2SO_4 is diprotic acid and $HC_2H_3O_2$ is monoprotic acid

112 TTCE

113 TTCE

114 FT

115 TTCE

[Part C]

26 (D)

27 (E)

28 (C)

29 (A)

30 (B)

31 (D)

32 (D)

33 (B)

34 (B)

35 (C)

36 (D)

37 (D)

38 (C)

39 (D)

40 (E)

41 (C)

42 (B)

43 (E)

44 (A)

45 (E)

46 (D)

At the standard state($0°C$, 1atm), the density of the gas is molecular weight (g)/22.4 (L), and the average velocity of gas molecules at the same temperature is inversely proportional to the square root of the molecular weight.

47 (E)

48 (E)

49 (C)

50 (D)

51 (C)

52 (B)

53 (A)

54 (C)

55 (D)

56 (B)

57 (E)

58 (A)

59 (B)

60 (E)

61 (D)

62 (A)

63 (D)

64 (B)

65 (C)

66 (E)

67 (E)

68 (C)

69 (A)

The normal boiling point is the temperature at which the vapor pressure of the liquid becomes 1atm.

70 (D)

MEMO

SAT II Chemistry

초판 발행 2021년 1월 8일

저자	임성택
발행인	최영민
발행처	헤르몬하우스
주소	경기도 파주시 신촌2로 24
전화	031-8071-0088
팩스	031 -942 - 8688
전자우편	hermonh@naver.com
출판등록	2015년 3월 27일
등록번호	제406-2015-31호
정가	23,000원
ISBN	979-11-91188-17-2 (53430)

• 헤르몬하우스는 피앤피북의 임프린트 출판사입니다.
• 이 책의 어느 부분도 저작권자나 발행인의 승인 없이 무단 복제하여 이용할 수 없습니다.
• 파본 및 낙장은 구입하신 서점에서 교환하여 드립니다.